SOME
BUILDING CONTRACT
PROBLEMS

SOME BUILDING CONTRACT PROBLEMS

VINCENT POWELL-SMITH
LLM, DLitt, FCIArb

BSP PROFESSIONAL BOOKS
OXFORD LONDON EDINBURGH
BOSTON MELBOURNE

Copyright © Ingramlight Properties Ltd
1989

First published 1989

British Library
Cataloguing in Publication Data

Powell-Smith, Vincent
    Some building contract problems.
    1. Great Britain. Buildings. Construction.
    Contracts. Law
    I. Title
    344.103'7869

    ISBN 0-632-02581-6

BSP Professional Books
A division of Blackwell Scientific
    Publications Ltd
Editorial Offices:
Osney Mead, Oxford OX2 0EL
    (Orders: Tel. 0865 240201)
8 John Street, London WC1N 2ES
23 Ainslie Place, Edinburgh EH3 6AJ
3 Cambridge Center, Suite 208, Cambridge
    MA 02142, USA
107 Barry Street, Carlton, Victoria 3053,
    Australia

Set by DP Photosetting, Aylesbury, Bucks
Printed and bound in Great Britain by
Mackays of Chatham PLC, Chatham, Kent

Contents

Preface

The last 20 years have seen the rapid development of construction law and, as legal correspondent of *Contract Journal* since 1974, I have tried (sometimes unsuccessfully) to keep abreast of the changes in my weekly column angled at those working in the construction industry. Inevitably, over that time period the legal issues have fallen into a pattern and there have been many recurring themes. The legal trends have sometimes departed from the industry's expectations – I instance, for example, the recent and welcome narrowing of negligence liability. Many important issues affecting the industry have been decided by the judiciary, and a surprising number of construction industry disputes now end in arbitration or litigation.

This book is a discursive discussion of *some* problem areas in construction law – problem areas at least so far as those at the sharp end are concerned. The selection is a personal one reflecting those issues which have been recurring themes in my column. Surprisingly, many people have filed away the *CJ* articles for reference and I have found what I have written quoted back at me many years afterwards, even when the law has changed.

The book is not a complete or balanced coverage of building contract problems. My approach has been eclectic and selective and I am conscious of the gaps. It is a collection of essays based on what I have written over the years. I hope that it will stimulate discussion and be of interest and help to contractors and sub-contractors as well as to architects and quantity surveyors, if only as highlighting areas of actual or potential difficulty and emphasising the changing approach of the law. In some cases the answers given to the industry's questions by Her Majesty's judges have caused consternation even though, in many instances, the issue has resolved itself into one of what the disputants actually agreed. Despite the ready availability of specialist series of law reports, legal columns in the technical press, numerous books and various regular digests of legal developments, much of the industry's business appears to be carried on in happy ignorance of the law, with the result that the industry is a major benefactor of the legal and consultancy professions.

I am grateful to successive editors of *Contract Journal* – John Osborne, Alex Catto and Jerry Gosney – for allowing me the freedom of their pages, and I owe a special debt to George Battley, its long-standing assistant editor who has been more helpful to me in my writing than he can ever know. *Contract Journal* readers have stimulated me by providing ideas or putting other points of view, and so this book is dedicated to all those associated with the magazine.

Once again, I owe thanks to Julia Burden of the publishers who readily took my suggestion for the book on board and then saw a very difficult manuscript through the press. Even in these days of high technology, publishing is essentially a personal business. The production of a book requires not only an author but a hard working and inspiring commissioning editor: Julia Burden has fulfilled that role for me for a great many years and I am glad to acknowledge her help and friendship.

I am also grateful to Dr John Parris for his helpful comments and criticisms, although I have not been able to accept them all. The text remains my sole responsibility.

The law is stated from sources available to me here on 1 December 1988.

Cerro das Covas   *Vincent Powell-Smith*
Portugal

List of Abbreviations

ACA Form	Association of Consultant Architects Form of Building Agreement, Second Edition, 1984
BEAMA RC Conditions	Federation of British Electrotechnical and Allied Manufacturers Associations Conditions of Contract for Reconstruction, Modification or Repair of Plant and Equipment
DOM/1	Standard Form of Domestic Sub-contract for use with JCT 80
FIDIC Form	Fédération Internationale des Ingénieurs Conseils Conditions of Contract for Works of Civil Engineering Construction
GC/Works/1	General Conditions of Government Contracts for Building and Civil Engineering Works, Edition 2, 1977
ICE Conditions	Institution of Civil Engineers Conditions of Contract, Fifth Edition
IFC 84	JCT Intermediate Form of Building Contract 1984
JCT	Joint Contracts Tribunal for the Standard Form of Building Contract
JCT 63	JCT Standard Form of Building Contract 1963
JCT 80	JCT Standard Form of Building Contract 1980
MW80	JCT Agreement for Minor Building Works 1980
NAM/T	JCT Form of Tender and Agreement for Named Sub-contractors under IFC 84
NFBTE/FASS 'Green Form'	Standard Form of Sub-contract for Nominated Sub-contractors for use with JCT 63
NSC1	JCT Nominated Sub-contract Tender and Agreement for use with JCT 80
NSC4	JCT Standard Form of Sub-contract for Nominated Sub-contractors 1980

Chapter 1

Contract Formation

How rough is an estimate?

Architects, quantity surveyors and other professionals asked to give written or verbal estimates of the likely cost of proposed works must make sure that the exact basis of the estimate is stated clearly. They must say whether or not the estimate includes for any possible increase in cost and the extent to which it does so. Failure to do so can amount to professional negligence.

This important point emerged from the Court of Appeal's decision in *Nye Saunders & Partners* v. *Alan E. Bristow* (1987) where the architect sued for professional fees amounting to £15 581.59. Bristow defended on the basis that the architectural firm had failed in its duty to take reasonable care in providing a reliable approximate estimate of the cost of renovating his mansion in Surrey in February 1974. In particular, it had failed to draw his attention to the fact that inflation would inevitably increase the estimated costs beyond his budget.

Nye Saunders was retained by Bristow in 1973 to prepare and submit a planning application for the renovation of his home. He said he had about £250 000 to spend. Before planning permission was granted, Bristow asked the architect for a written estimate of the cost. Prudently, Nye Saunders consulted a quantity surveyor, and wrote to Bristow giving

> 'an approximate estimate [and] in order to give you as accurate a figure as possible I consulted my quantity surveyor and I enclose his letter and schedule of costs'

which amounted to £238 000.

Work began on the detailed plans and then disaster struck. The architect advised Bristow that:

> 'there is no possibility that the tenders which we are likely to receive will be within a reasonable bracket of my original, approximate estimate.'

At a subsequent meeting, the client was given a revised estimate totalling some £440 000. He cancelled the project and in effect terminated the architect's engagement, adding: 'I do not know how you could be nearly 100% out in your cost estimates.'

Effectively, the allegation against the architect was that in giving the estimate of £238 000 in early 1974 and then revealing that the ultimate cost would almost double – in particular because of inflation – it had failed in its duty of care. In cases like this, the courts approach the matter on the basis of considering whether there was evidence that at the time of the alleged negligence a responsible body of architects would have acted as the defendant had done in carrying out his duties.

The trial judge heard expert evidence on this point and ruled that the architect was responsible. He said:

'I find that there was not a practice accepted as proper by a responsible body of architects in February 1974 that no warning as to inflation need be given when providing an approximation of the cost of the then current outline proposals.'

The Court of Appeal agreed. The fact that the architect had consulted a quantity surveyor did not avail him. The architect was himself responsible for the estimate.

Lord Justice Stephen Brown said:

'Of course it was a very sensible and prudent step for [the architect] to take to consult a quantity surveyor who is an expert in computing costs. But in my judgment he cannot avoid responsibility for the fact that he did not draw the attention of his client to the fact that inflation was not taken into account. The duty rested fairly and squarely on [the architect] and it cannot be avoided by, as it were, seeking to move the responsibility on to [the quantity surveyor].'

The quantity surveyor had carried out his practice of costing at current prices, but what Bristow wanted to know – and asked the architect – was: What is it going to cost me?

Essentially, that disposed of the appeal, and the architect lost his case. The trial judge, however, following the approach of Lord Bridge in a medical negligence case (*Sidaway* v. *Governors of the Bethlem Royal Hospital* (1985)), gave an alternative basis for his decision. In light of the inflation figures shown by the evidence, he also found that:

'the disclosure of the risk of inflation upon the costs of the proposals was so obviously necessary to enable the client to make an informed decision that no reasonably prudent architect would fail to make it'.

The Court of Appeal did not expressly dissent from that proposition. Indeed, on the facts as disclosed it does seem that it was an obvious matter that inflation should be taken into account.

As the Court of Appeal noted, the term 'negligence' has a nasty ring, particularly when applied to a professional man, but ordinary carelessness is not negligence.

'This case concerns a duty which the law says is placed upon somebody who undertakes the responsibility of providing an answer such as Mr Nye had undertaken to do.'

The ruling affects contractors as well as the professional team. A small contractor, for example, asked to give an estimate for renovation or alteration works must make its basis plain. The estimator must bring his client's attention to everything which may affect the ultimate cost. All too often estimates are prepared on a very conservative basis.

The true position is accurately summarised in the commentary on the case in *Building Law Reports*:

'All estimates are "approximate" and most clients are unable to appreciate the variations of which the professional ... is presumably aware. This is particularly the case in relation to items for which prime costs or provisional sums are provided.'

Now that inflation is under control, that aspect of *Nye Saunders* v. *Bristow* is not vital. But the essential point is that the basis of all estimates must be accurately and fully stated so that the client can make an informed decision.

Contractors must also take care since a contractor's estimate to the client will usually be held to be a firm offer in law. Its unequivocal acceptance will result in a binding contract. Thus, in the old case of *Croshaw* v. *Pritchard and Renwick* (1899), the plaintiff invited a tender for some alteration work, enclosing a drawing and specification with his invitation. The defendant replied: 'Estimate – Our estimate to carry out the ... alterations to the above premises according to the drawing and specifications amounts to £1230'. The plaintiff replied accepting the estimate. It was held that the 'estimate' was an offer accepted by the plaintiff's reply and so there was a binding contract.

Letters of intent

As letters of intent are so common in the construction industry, it is surprising that there is so little case law about them. Most of the books ignore the subject completely or, where they touch upon it, refer to the case of *Turriff Construction Ltd* v. *Regalia Knitting Mills Ltd* (1971) which in many ways is an unsatisfactory case because, on the facts, there was held to be contractual liability.

In the normal course of things, a letter of intent merely expresses an intention to enter into a contract in the future. It creates no contractual liability in itself but the wording of the letter, and the facts of each case, may give rise to liability. The position was clarified by the High Court decision in *British Steel Corporation* v. *Cleveland Bridge and Engineering Co. Ltd* (1981), where a sub-contractor commenced work pursuant to a request to proceed immediately, pending the execution of a form of sub-contract. No sub-contract was entered into.

Mr Justice Robert Goff held that no contract had come into existence between the parties, despite the letter of intent, and the performance of the work by the sub-contractor. He held further that the sub-contractor was entitled to be paid on a *quantum meruit* basis ('as much as it is worth') and judgment was entered for the sub-contractor for £229 832.70.

BSC was the sub-contractor for the manufacture and supply of 137 cast steel nodes, to be used in the construction of a bank in Saudi Arabia.

The letter of intent said that it was Cleveland's intention 'to enter into a sub-contract with (BSC) for the supply and delivery of the' nodes. It asked BSC to 'proceed immediately with the works pending the preparation and issuing ... of the official form of sub-contract.'

A subsequent telex from Cleveland required the nodes to be manufactured in a particular sequence. The parties never agreed on the price or the delivery dates and Cleveland paid nothing for the delivered nodes. Subsequently, Cleveland put in a counterclaim for breach of contract by late delivery.

If there was no binding contract between the parties there was no basis for Cleveland's counterclaim and so it was important to BSC that the judge should find that they were entitled to be paid on a *quantum meruit* basis, as he did.

Mr Justice Robert Goff said that there was no hard and fast answer to the question whether a letter of intent would give rise to a binding agreement: everything depended on the circumstances of the case.

There were two legal possibilities contractually. First, there might be an ordinary executory contract, under which each party assumed reciprocal obligations. Second, there might be an 'if' contract, i.e., a contract under which A asks B to carry out certain work and promises B that, if he does so, he will receive performance (usually remuneration) in return. This was really no more than a standing offer.

Cleveland argued that the first type of contract had been formed here, but that argument could not be sustained.

The request to BSC to proceed was stated to be 'pending the preparation and issuing ... of the official form of sub-contract.' This indicated that the sub-contract was plainly in a state of negotiation. By starting work BSC had not bound itself to any contractual performance.

Much of the same difficulties applied in the second type of situation, which was certainly analytically possible, but had to be rejected because it would also provide a vehicle for certain contractual obligations relating to performance, such as implied terms as to the quality of the steel nodes.

'Neither party knew what the liabilities of the other would eventually be under the contract,' the judge said, including provisions relating to liquidated damages and exemption clauses. If a buyer asks a seller to commence work 'pending' the entering into of a formal contract, it is difficult to infer from the seller acting upon that request that he is assuming any responsibility for his performance, except the responsibility which will rest upon him under the terms of the contract which both parties confidently anticipate they will shortly enter into.

The judge said:

> 'It would be an extraordinary result, if by acting on such a request
> ... the seller were to assume an unlimited liability for his contractual
> performance, when he would never assume such liability under any
> contract which he entered into.'

Mr Justice Robert Goff said that the true analysis was that both parties expected a formal contract to come into being later. To expedite performance under that anticipated contract, Cleveland requested BSC to commence work, and BSC complied with that request. The anticipated contract was never entered into, and so the performance of the work was not referrable to any contract of which the terms could be ascertained. In such circumstances the law simply imposed an obligation on Cleveland to pay a reasonable sum for the work which had been done pursuant to its request.

Although there are difficulties in this approach, the decision accords with commercial realities and with the views of those who send and act upon letters of intent. Such letters usually express an intention to enter into a contract in the future, without creating contractual liability. But if acted upon, in the usual way, the person doing the work does so in anticipation that he will be paid.

Of course, the letter may itself deal with that question by negativing any liability; but that is not usual. Much depends on the text of the letter, the circumstances in which it was written, and the subsequent actions of the parties. If a formal contract is entered into, then of course its terms will have retrospective effect: *Trollope and Colls Ltd* v. *APC Ltd* (1962). It is only where no formal contract is executed that difficulties are likely to arise.

The greatest care is needed in these situations, but at least the *Cleveland Bridge* case shows that modern commercial judges take a realistic approach. Specialist sub-contractors undertaking preparatory work in like circumstances can take heart.

Paying the bill for errors

The unreported Court of Appeal decision in *Dudley Corporation* v. *Parsons and Morrin Ltd* (1959), April 8, (abstracted in *A Building Contract Casebook*, p. 41) is said by some writers to support the view that the contractor is bound by errors in his bill rates as regards both original and varied work.

In fact, the case does nothing of the sort because the arbitrator found that 'no sufficient evidence was brought before me that the price ... inserted ... was a mistake' and, as the judgments in the Court of Appeal make clear, the case turned purely on the interpretation of the contract documents.

The proposition is, however, a sound one, and pricing errors bind both parties: see, e.g. Keating's *Building Contracts*, 4th edition, p. 315, discussing the position under JCT 63. Indeed, in a case not cited to the court (*Higgins* v. *Northampton Corporation* (1927)) Mr Justice Romer held that a contractor who makes a unilateral mistake in his tender price cannot claim to have the contract set aside or rectified on the ground that he was labouring under a mistake. The *Dudley Corporation* case is, however, of general interest for its facts and the ruling and the author is indebted to the present chief executive of Dudley Corporation for the transcript.

The contract was in Royal Institute of British Architects' (RIBA) 1939 form, but the relevant provisions are similar in JCT 80. The contractor priced an item for '750 yards cube. Extra only for excavating in rock ... (Provisional) 2s £75.'

A fair and reasonable price would have been £2 a cubic yard. The contractor actually excavated 2230 cubic yards of rock. The architect valued the excavation at 750 yards at 2*s*, and the balance of 1480 cubic yards at £2.

The corporation disputed this valuation and contended that the contractors were only entitled to be paid an extra of 2*s* a cubic yard for all excavation in rock over the quantity provisionally allowed for excavation in the bills. The contractors argued that the valuation was for the right amount.

The Court of Appeal, reversing the trial judge, held that the contractors were entitled to 2*s* a cubic yard for the whole quantity excavated on the basis that this is what the Bills said. In his judgment, Lord Justice Hodson emphasised the point that the bills had been carefully and specially prepared and went through several of the arbitrator's findings of fact.

One of the most important of these was that

> 'having regard to the depths of the various excavations given in the bills ... and drawings and the evidence as to the nature and condition of the site', it could not be a reasonable assumption that if rock was encountered it would be found only in the area of 'the heating basement and nowhere else on the site'.

The contractor contended that the words 'extra only' referred to the heating basement.

Lord Justice Romer's judgment is short and interesting. He concluded that the words referred to all the previous items under the bill heading of 'Excavator' and he did so for two reasons. First, the item 'extra only' was in absolutely general terms and covered all the numerous items of excavation which preceded it.

Secondly:

> 'The draughtsman of the bill of quantities has repeatedly used the phrase "do. and do." both in the excavator and drainlayer sections of the bill ... when he intends to connect an item with its immediate predecessor. He has most significantly omitted to do so in the case of the "Extra only" item. Had he intended to confine the application of this item to the three [heating basement items] ... it is almost impossible to suppose that he would not have adopted the practice which characterises the rest of the document ... which ... was drawn with some care and precision.'

Thus he felt that the extra was of general application to all the preceding items.

Lord Justice Pearce made an interesting additional point which is the crux of the matter in practical terms. He said:

> 'The actual financial result should not affect one's view of the construction of the words. Naturally one sympathises with the contractor in the circumstances, but one must assume that he chose to take the risk of greatly under-pricing an item which might not arise, whereby he lowered his tender by £1,425. He may well have thought it worth while to take that risk in order to increase his chances of securing the contract.'

Dudley Corporation v. *Parson and Morrin Ltd* is clearly not relevant to questions of errors in pricing, on which there is little building case law. General contract cases establish that in a lump sum contract (as is JCT) errors not discovered by the employer or the architect before acceptance of the tender bind both parties. The orthodox view is that this applies to extra quantities or varied work as well.

The JCT 80 provision for the correction of errors (clause 2.2.2.2) refers only to items and not to prices, and such loss cannot be recovered under JCT 80, by invoking clause 13.5.6 with its reference to:

> 'valuation of any work or liabilities directly associated with a variation (which) cannot reasonably be effected in the valuation by the application of clause 13.5.1 to 13.5.5 (then) a fair valuation thereof shall be made.'

If the rates are too high, of course, then the contractor gets the benefit, but the practical answer is that the contractors must scrutinise their rates very carefully when putting in the tender because those rates will apply to variations as well as to the original contract sum.

There is a difference between what the contract says and what some people think it should say and, certainly under JCT contracts, it seems that both parties are stuck with the erroneous rate.

The position is, of course, different if the employer or the architect discovers the error before acceptance of tender and tries to take advantage of it and, in those circumstances, the court might rectify the mistake: see *A. Roberts and Co. Ltd* v. *Leicestershire County Council* (1961). But rectification is a remedy which is rarely ordered.

The Dudley Corporation case is a warning to draughtsmen of bills of quantities as well as to contractors; and it has often been said that hard cases make good law. The principles there laid down apply to all modern forms of contract. Only the amounts have changed.

The view that a contractor is bound by pricing errors in the contract bills, where the contract is let under JCT 80 terms, not only as regards the original quantities, but also in respect of varied work or additional quantities, is often challenged by quantity surveyors, even though *Higgins* v. *Northampton Corporation* is decisive on the question of the contractor being bound by his tender error.

Quantity surveyors often say:

'The equitable answer is that a contractor should not suffer loss as a result of a variation, unless it is for reasons under his control'.

An alternative approach has been mooted in discussion. It is said that:

'If there is a major addition to the contract, the architect has a choice between ordering a variation and arranging a new contract. Under JCT terms the contractor must execute the variation. If his existing rates are underpriced, surely the employer is bound to pay the fair price?'

The short answer to both questions is to be found in the terms of the contract – and the contract wording is quite clear. True, the contract does not say anything expressly about the application of erroneous bill rates in valuing variations. This is a matter of general law, and the architect (or quantity surveyor) cannot dispense with the contract requirements. The provisions in JCT 80, clause 2.2. dealing with discrepancies refer only to *items* and not to prices; the JCT forms contain no mechanism for the correction of errors in pricing, as already noted.

Rates in the contract bills, erroneous or otherwise, apply to the valuation of variations. Keating's *Building Contracts*, 4th edition, p. 315, quotes an interesting example: A contractor erroneously inserts a unit price of £5 instead of £50 and gives a total of £1000, excluding preliminaries, pc and provisional sums, whereas if £50 had been written down it would have been £1045.

If a further five units of the item are required by way of variation, must the contractor carry them out for 5 × £5, thereby multiplying the effect of his error?

Keating says:

'It is thought, having regard to clause 13, the contractor is bound by his £5 for the purposes of valuation under clauses 11(4)(a) and (b) ... the contractor cannot recover the difference between £5 and his intended price of £50 per unit as "loss or expense" under clause 11 (6).' [of JCT 63.]

Walker-Smith and Close in *The Standard Forms of Building Contract*, p. 43, observe that a distinction:

'must be drawn between an instruction which causes direct loss, which is the subject of reimbursement under sub-clause (4), albeit this may be less than total compensation, and a direct loss which is not properly covered by sub-clause (4) at all,'

and say that clause 11(6) cannot be used to supplement any deficiencies.

Both writers are commenting on JCT 63, but the position is identical under JCT 80, clause 26.2.7 of which applies only to *items* not coming within clause 13.4.5 and 6.

Keating's reference to JCT 63, clause 13 (JCT 80, clause 14) is important: the contractor has undertaken to carry out the works as varied, variations being envisaged by the contract itself and:

'... any error whether of arithmetic or not in the computation of the contract sum shall be deemed to have been accepted by the parties.'

The re-run of the clause 11(6) argument under JCT 80 is by reference to clause 13.5.6 which envisages times when 'a fair valuation' shall be made, but only where the quantity surveyor decides that 'the valuation of any work or liabilities directly associated with a variation cannot reasonably be effected' by the application of the general rules.

Underpriced work cannot be valued under JCT 63, clause 11(4)(b) (JCT 80, clause 13.5.1.2) as is sometimes suggested. This is limited to the valuation of work of a different character under different conditions to those envisaged in the original contract; and it would need very clear words indeed to overturn the general law of contract.

The contrary view put forward is based on the observations of Wallace, *Building and Civil Engineering Standard Forms*, p. 25, where it is submitted that JCT 63, clause 12(2), by its wording suggests that both clause 11(6) and the 'fair valuation' provisions of clause 11(4):

'can theoretically be invoked by the contractor in a case where there is an error in the bills of an unusual character ... [but] only errors of the most abnormal kind and magnitude could possibly justify a departure from the bill prices.'

In any event, as the commentary makes clear, this does not apply to pricing errors.

In short, the contract requires all variations to be valued at bill rates unless the circumstances differ from those implicit in the tender documents. Such differences must be fairly valued, but the bill rates provide the basis. The position is similar under JCT 80.

So far as varied work is concerned, the architect's only choice when requiring additional work is to consider whether it can properly be ordered as a variation, as defined in JCT 80, clause 13.1, which states that no variation shall vitiate the contract.

There is, of course, a limit to the extent and nature of variations which can be ordered, because it is only the works as set out and described in the contract documents that may be altered or modified. The architect's power does not extend to ordering any additions, etc., which would require the contractor to execute work clearly not contemplated by the original contract. He cannot substantially alter the nature of the works – but it is only in those circumstances that a 'new contract' would be necessary.

No doubt claims have been made and met on the basis of the alternative views put forward but this does not make them correct.

The 'battle of forms'

Not all construction work is done on standard form terms. Numerous variant contract forms are used, particularly by specialist sub-contractors. Happily, most of them are based on one or other of the accepted standard forms, none of which can be described as the perfect document.

In the cut and thrust of business, each contracting party with his own standard terms tries hard to ensure that his terms will prevail, even where they conflict with those of the opposition. The resulting legal tangle is known to lawyers as 'the battle of forms'.

This is where the parties attempt to contract by exchanging orders and acknowledgements, each of which purports to be on the usual terms of the party putting it forward. It is often uncertain who is going to win the battle, and there are differing views as to the rules of war. The battle is fought on classical lines.

In general, the correct approach is to see whether one party has by words or conduct accepted the other's terms. If there has been no 'acceptance' there may well be no contract!

The major case is *Butler Machine Tool Co. Ltd.* v. *Ex-Cell-O-Corporation Ltd*, (1979), and involved the sale of machine tools valued at £75 500. Sellers quoted to sell the tools at that price, delivery ten months forward. The offer was stated to be subject to certain terms and conditions, one of which read:

> 'All orders are accepted only upon and subject to the terms set out in our quotation and the following conditions. These ... shall prevail over any terms and conditions in the buyer's order.'

The seller's terms included a price variation clause.

The buyer's acceptance – in the form of a standard purchase order – purported to incorporate its own contract conditions which were materially different from those put forward by the sellers. In particular, the buyer's conditions were on a fixed-price basis. At the foot of the buyer's order was a tear-off acknowledgement slip saying: 'We accept your order on the terms and conditions stated thereon.'

The sellers completed and returned this acknowledgement slip along with a covering letter saying that the buyer's order was being entered in accordance with the seller's quotation.

When the machine tools were delivered, the seller claimed a price increase of nearly £3000, and this the buyer refused to pay. The seller started legal proceedings claiming entitlement to a price increase under the variation clause. The buyer contended that the contract had been concluded on its terms and was thus a fixed-price contract.

The Court of Appeal's decision makes interesting reading. The court ruled that the sale was on buyer's terms, reversing the decision of the High Court judge who had found in the seller's favour on the basis that its terms were to prevail as had been stipulated in the opening offer.

The three Appeal judges arrived at the answer by analysing the problem on the basis of 'offer and acceptance.' On this footing, the buyer's order was a counter-offer which destroyed the offer in the seller's quotation. By completing and returning the acknowledgement of order, the seller had accepted the counter-offer on the buyer's terms. Accordingly, the seller could not claim an increased price under its own dead terms.

The covering letter referring to the seller's original quote was irrelevant in their lordships' view; it was to be read as merely referring to the price and identity of the machine tools and did not incorporate the seller's terms back into the contract.

Lord Justice Lawton and Lord Justice Bridge were content to decide the parties' fate 'in accordance with set rules' or 'the classical doctrine that a counter-offer amounts to a rejection of an offer and puts an end to the effect of the offer'. That is the traditional analysis and the one which most lawyers use when considering the effect of conflicting terms and conditions.

The seller returned the acknowledgement slip – the last document in time – and this said: 'We accept your order on the terms and conditions stated thereon' and there was an end of the matter. As Lord Justice Bridge said:

> 'At that moment in time, there was a complete contract in existence, and the parties were *ad idem* as to the terms of the contract embodied in the buyer's order.'

Lord Denning MR, reached the same conclusion, but he went further. He thought that the documents comprised in a 'battle of forms' were to be considered as a whole to discover the agreement of the parties. This should be done by reasonable implication if the conflicting terms and conditions of both parties were irreconcilable. On that basis, the acknowledgement of the order was the decisive document since it made it clear that the contract was to be on the buyer's and not the seller's terms.

On Lord Denning's analysis, the contract comes into being as soon as the last of the conflicting forms is sent and received without objection being taken to it.

> 'The difficulty is to decide which form, or which part of which form, is a term or condition of the contract. In some cases the battle is won by the man who fires the last shot. He is the man who puts forward the latest terms and conditions: and if they are not objected to by the other party, he may be taken to have agreed to them.'

That is the general position.

Lord Denning departed from orthodoxy – and probably went too far – in suggesting that as a long-shot:

> 'the terms and conditions of both parties are to be construed together ... If the differences are irreconcilable, so they are mutually contradictory, then the conflicting terms may have to be scrapped and replaced by a reasonable implication.'

That seems to me to come perilously close to the court re-writing the contract for the parties.

In practical terms, those wishing to contract on their own conditions must be meticulous in contract documentation. Don't return any acknowledgement slip – and if the other side persists in sending 'purchase orders' with terms which conflict with yours, make sure that the last letter – and prevailing conditions – are yours.

Many standard contract forms attempt to do this: see, for example, clause 3 of the Model Conditions of Plant Hire and various standard forms of Purchase Order. The 'last shot' doctrine is the one which will usually be applied, and it is doubtful whether Lord Denning's 'reasonable implication' theory will find acceptance, as a later case shows.

That decision is a construction industry case: *Chichester Joinery Ltd* v. *John Mowlem & Co. Plc* (1987) where Chichester sued Mowlem for money allegedly due for joinery which they had manufactured and supplied to Mowlem for incorporation in two buildings at Royal Holloway College. Chichester's main contention was that the contract was made on 30 April 1985 and subject to its standard conditions, although it changed its ground slightly. Mowlem said that the agreement was made on 14 March 1985 and subject to the terms and conditions of its Purchase Order.

His Honour Judge James Fox-Andrews QC, official referee, started his judgment with a warning which is of general application.

'This case highlights the risks facing parties who seek to impose their own respective conditions, rather than using some well-established standard form of contract commonly used for the supply of goods to a main contractor for incorporation in construction work. Unless such a form of contract is used, the Courts may be faced (as here) with what is essentially an artificial state of affairs. It is not uncommon for the parties to be advised to ensure that whenever a quotation, an offer, an order, an acknowledgement of acceptance, etc. is sent, that a copy of the conditions is endorsed on or annexed to the document. It often occurs that the typed-in portions do not correspond with the printed description of the document'.

Chichester's original quotation to Mowlem was submitted to Mowlem in November 1984 under cover of a printed letter dated 31 August 1984. A copy of its standard conditions was sent with the letter, which said that:

'this quotation is based on our standard conditions of sale overleaf, to which your attention is drawn. These conditions can only be varied by our agreement in writing'.

In January 1985, Mowlem sent out to Chichester and others an enquiry document, referring to printed conditions which, so the judge found, were not necessarily the same as those on the subsequent Purchase Order. A telephone discussion took place between the parties, when Chichester confirmed the rates quoted on 31 August 1984 and agreement was reached that there should be a cash discount of 5%. It was also agreed that February 1985 would be the base date for NEDO formula calculations. Chichester then confirmed in writing that they would do the work in accordance with their original quotation and the other agreed terms.

Mowlem's Purchase Order was sent on 14 March 1985. It referred to the order being placed subject to printed terms on its reverse, and stated that 'Your quotation, if any, is taken only as a basis for pricing'.

There were significant differences between the parties' sets of conditions. A note to Mowlem's Purchase Order said that there was to be a 10% non-adjustable element, to which there had been no previous reference.

The judge rejected Mowlem's argument that this was an insignificant term. 'I do not accept that. It is a term which can be of considerable significance to a supplier', he said.

He went on to hold that the Purchase Order was clearly not an acceptance of Chichester's offer. It was a counter-offer, and killed Chichester's offer. It was argued that Chichester's subsequent conduct in preparing to manufacture the joinery was acceptance by conduct. The judge rejected that argument as well. The evidence fell far short of what was necessary to establish acceptance by conduct. Chichester neither orally nor in writing either within seven days, or at all, accepted Mowlem's before 30 April 1985 – a substantial time before any deliveries were made – when they sent a printed 'Acknowledgement of order' which incorporated identical conditions to those on their original 1984 quotation. Although headed 'acknowledgement' some of its wording was more consistent with acceptance than with mere acknowledgement.

The judge was unable to spell out from the relevant facts any intention by Chichester to be bound by Mowlem's conditions. The facts were significantly different from those in *Butler Machine Tool Co. Ltd* v. *Ex-Cell-O-Corporation Ltd* (1979) where an acknowledgement slip was completed and returned, thus accepting a counter-offer. In this case the reverse was true: Chichester's 'acknowledgement' was the counter-offer, which killed Mowlem's offer of 14 March.

This left one question to be determined. Did Mowlem's subsequent acceptance of delivery of the joinery on site amount to an 'acceptance' of Chichester's counter-offer, so that the contract was subject to Chichester's conditions?

Judge Fox-Andrews said:

'These cases are never easy. After some hesitation I have reached the conclusion that Mowlem did accept Chichester's terms. Mowlem at an earlier stage was specifically offering to treat a contract as complete if, on joinery being delivered to site, then was accepted by them'.

He therefore decided the point in Chichester's favour.

How much was at stake in the case does not appear from the judgment which was on a preliminary issue of law. But the case is illustrative of the practical difficulties of the 'battle of forms' in a construction industry context. On slightly different facts, things could have gone the other way!

More than a mere formality

Contracting organisations are often very lax when it comes to completing contractual formalities. In many cases large-scale projects have been well-nigh completed before the actual contract documents were signed.

This, and related problems, are very surprising to the legal mind. There is a large number of cases which illustrate the principles involved and the attitude of the courts to what is no more than poor contract administration.

If a formal contract is entered into after work has been started, then in principle its terms will have retrospective effect so as to cover the work done before contract. This follows from *Trollope and Colls Ltd* v. *Atomic Power Constructions Ltd* (1962) where there were very lengthy negotiations which continued while work went on.

Work had begun in June 1959, but the contract was not actually concluded until 11 April 1960. As Mr Justice Megaw pointed out, if a contract could not have retrospective effect, 'there would be many important mercantile contracts which would, no doubt to the consternation of the parties, be nullities.'

However, the court will not create a contract for the parties, and it may be that the parties may reach agreement on broad matters of principle but leave important points unsettled, so that their agreement is incomplete. In that case, there is no contract: *May and Butcher* v. *The King* (1934).

Equally, a mere 'contract to negotiate' is too uncertain to be enforced, as is graphically illustrated by the construction industry case of *Courtney and Fairburn Ltd* v. *Tolaini Bros Hotels Ltd* (1974).

There, developers approached a contractor, and agreed that, if he arranged the necessary finance for a motel project, they would be prepared to instruct their:

> 'quantity surveyor to negotiate fair and reasonable contract sums in respect of each of the three projects as they arise ... based upon agreed estimates of the net cost of the work and general overheads with a margin of 5 *per cent*.'

Despite the fact that the contractor arranged the necessary finance, the developers placed the contracts elsewhere.

The trial judge held that there was a binding and enforceable contract, but was overruled by the Court of Appeal. A mere 'contract to negotiate' is too uncertain to be enforced; there was no binding contract because there was no agreement on the price or any method by which it was to be calculated. The use of the word 'negotiate' seems to have been one of the critical factors; it is too vague to be enforceable.

This case has been criticised by many leading contract lawyers, and in Scottish law it has been held (in *R. and J. Dempster Ltd* v. *Motherwell Bridge and Engineering Co. Ltd* (1964)) that there was a valid contract for civil engineering work based on an exchange of letters which used the words 'the prices to be mutually settled at a later and appropriate date'.

Moreover, it is clear that there can be a 'contract to make a contract', although Lord Denning appeared to think otherwise in the *Tolaini* case. For example, in the family law case of *Morton* v. *Morton* (1942) an undertaking was signed under the headnote of 'heads of agreement' to 'enter into a separation deed containing the following clauses' which were then summarised. This was held to be a binding contract. Much will, therefore, depend on the words used and on the surrounding circumstances.

Although the courts will not make a contract for the parties, they will seek to uphold bargains whenever possible by seeking out the true presumed intention. This was recognised by Lord Wright in *Hillas and Co. Ltd* v. *Arcos Ltd* (1932), when he spoke of the habit of business people who record:

> 'the most important agreements in crude and summary fashion; and modes of expression sufficient and clear to them in the course of their business may appear to those unfamiliar with the business far from clear and precise.'

The modern judicial approach is well-illustrated by *Modern Building Wales Ltd* v. *Limmer and Trinidad Co. Ltd* (1975) where there was an exchange of correspondence between a main contractor and a nominated subcontractor, placing an order 'in full accordance with the appropriate form for nominated sub-contractors (RIBA 1965 edition) ... all as your quotation.'

The Court of Appeal heard expert evidence to the effect that there was no RIBA 1965 form for nominated sub-contractors, but that 'the appropriate form for nominated sub-contractors' meant the old NFBTE/FASS 'green form'. Accordingly, the whole of that document was held to be incorporated into the contract between the parties.

Loose language in a sub-contracting situation was also considered by the Court of Appeal in *Brightside Kilpatrick Engineering Services* v. *Mitchell Construction (1973) Ltd* (1975), where the order form referred to 'the conditions applicable to the sub-contract ... shall be those embodied in the RIBA as above agreement' meaning JCT 63. Once again the Court held that the terms of the 'green form' applied, but for a different reason.

Lord Justice Buckley took the view that the use of this phrase meant that the sub-contractual relationship should be such as to be consistent with all those main contract terms specifically dealing with nominated sub-contractors so that there should be no conflict between the main and sub-contract.

In both these cases the courts reached a realistic and commercially workable solution, but at what price? In neither instance did the parties enter into a formal sub-contract, and all the costs of the litigation and its inevitable delays could have been avoided had the parties turned their minds to the situation and got down to the paperwork.

A just result was in fact achieved but in any of the cases discussed in this article, the decision might equally have gone the other way.

The law is not always as certain as lawyers sometimes suggest. All those involved in contract administration would do well to check up on their own procedures and tighten up where necessary.

The vagaries of the written word

If experience is anything to go by, far too few people read contracts before signing them or, if they do, they fail to understand them. This may be the fault of the lawyers responsible for drafting the documents but, in many cases, the reader makes assumptions which have no basis in reality. An amendment to a standard clause – or a specially-drafted clause – can significantly affect the risk and hence the cost.

For example, since the Court of Appeal's ruling in *Rees & Kirby Ltd* v. *Swansea City Council* (1985), a number of local authorities have been amending JCT 80, clause 26, so as to provide that the term 'direct loss and/or expense' shall *not* include or cover 'interest' or 'finance charges'. And surprisingly some contractors have signed contracts in that amended form. True, if JCT 80, clause 26, is correctly operated, the financial burden element in any claim is likely to be small – in contrast with the post-completion claims commonly made under JCT 63. But to accept amendments of this kind is to court disaster.

It is not only contractors who are at fault. Architects and quantity surveyors responsible for preparing and issuing contract documents are often slip-shod in their approach. Sometimes they fail to delete alternative and conflicting clauses or, more commonly, vital information is omitted from the main contract Appendix or in the other contract documentation. This laxity never ceases to surprise me – contractors must check that all is in order before committing themselves.

In many cases, of course, no formal contract is ever executed. This matters not one whit provided there has been a valid acceptance of the contractor's tender – assuming, of course, that it is in law a firm offer. If the contract is signed later, then its terms will have retrospective effect provided that the work done is clearly referable to the works described in the executed contract: *Trollope & Colls Ltd* v. *Atomic Power Constructions Ltd* (1962).

Another problem is the use of a totally inappropriate standard form for the project. The Joint Contracts Tribunal's series of standard forms has been specially designed to cover a wide range of works and for the most part they are backed up by suitable forms of sub-contract. The Minor Works Form (MW 80) is not designed for use where the project is valued at £1.5 million and with a contract period of 52 weeks and there are numerous specialist sub-contractors. But it has been used in such a project, presumably because the employer's team thought that it was short and simple.

Ideally, the contract should be tailor-made for the project in hand but, since this is not possible, the architect must seek the most suitable standard form, and the JCT makes a wide choice available to him. Should the employer wish to deviate or embellish the conditions in any way, it is advisable to take legal advice on the possible legal effect of the changes before proceeding.

Contractors faced with unusual or amended conditions should do likewise. 'For the sake of a nail, the shoe was lost.' In fact, from every point of view, standard form contracts should not be amended at all, with certain minor exceptions. For example, I would always urge the deletion of the JCT clauses which provide that nothing in the bills (or specification) shall override or modify or affect in any way the printed conditions. This is a nonsensical provision which has given rise to a great deal of litigation, and reverses the ordinary and sensible rule of interpretation under which specially prepared terms would prevail over the printed ones in case of conflict: *English Industrial Estates Corporation* v. *George Wimpey & Co. Ltd* (1973).

Even after the contract is executed, the problems are not at an end. The standard contracts all provide a clear procedural and practical guide as to what should happen; for example, clause 2.3 of JCT 80, dealing with discrepancies in and divergencies between the Contract Drawings, Contract Bills. Architect's Instructions and so on.

This does *not* impose on the contractor a positive duty to search for discrepancies or divergencies. The governing word is 'if'. But a wise contractor should take precautions because of the practical difficulties involved. The problem is practical and not legal, since the principal difficulty will be in securing payment from the quantity surveyor if the contractor has proceeded without obtaining an instruction under clause 2.3. Because of the wording of the clause, there is certainly no danger of the courts holding that there is an implied term that the contractor should seek out obvious errors, but prudence dictates that he should do so. The man on site is the person in the hot seat.

Indeed, communications on site (or lack of them) are the root cause of many of the problems put before me. Clause 10 of JCT 80 requires the contractor to keep a competent person-in-charge *constantly* on the works. He is the contractor's agent for the purpose of receiving instructions. His role is of especial importance in connection with oral instructions. Although the contract says that these 'shall be of no immediate effect', they are often of an urgent nature.

The competent person-in-charge may well be the contractor's site agent. It is vital for him to know and apply the provisions for written confirmation of oral instructions and, indeed, anything agreed verbally should be confirmed in writing immediately.

In a disputes situation, the person with the best and most accurate records is off to a head start.

Not quite by the book

Local and other public authority standing orders often lay down rules which must be observed in the making of contracts and so on, and questions are sometimes asked as to the effect of non-compliance by councils with standing orders. The answer is that mere non-compliance with standing orders does not affect the contractor who is protected if the internal contractual procedures are observed.

This follows from section 135(2) of the Local Government Act 1972 and its 1933 predecessor. That there are pitfalls for contractors, however, is well-illustrated by the decision of the Court of Appeal in *North West Leicestershire District Council* v. *East Midlands Housing Association* (1981) which presents a cautionary tale.

Shortly before local government re-organisation in 1974, the council's predecessors negotiated with the housing association for the building of three blocks of flats. In December 1973 they accepted the association's fixed price tender, and a draft contract in JCT 63 form was sent to them, heavily amended. At the last council meeting before re-organisation it was resolved that the chairman be authorised to deal with any urgent matters during the period prior to re-organisation.

On 18 March 1984 it was agreed between the clerk of the council and the association that the contract should be on a fluctuating basis, and the same day the contract was signed and sealed incorporating an amendment to that effect. The contract was signed by the chairman and the clerk but it did not comply with standing orders because no resolution authorising its sealing had been passed and the two witnesses were not present when it was sealed.

Work proceeded and interim certificates totalling £550 999 were paid by the council. The original fixed price tender had been for £468 985.36 and that was the offer which the former council had resolved to accept. The district auditor came on the scene and took the view that the contract was a fixed price one. The council brought an action claiming moneys overpaid, arguing that they had never intended to make or authorised the making of any contract other than on a fixed price basis. The association counter-claimed for £40 988.88 due on the final account on the basis that fluctuations applied. The trial judge, Mr Justice Swanwick, dismissed the council's claim and gave judgment for the association, holding that the fluctuation clause was a term of the contract.

The Court of Appeal reversed this ruling, but it was a pyrrhic victory for the council because in the event they recovered no money, while the association was left with the option of bringing a quasi-contractual claim.

The reasoning of the majority of the Court of Appeal is interesting. There were two points at issue. The first was whether the council's clerk had the authority to agree to the amendment. The terms of the council's resolution did not confer actual authority on him, but did he have ostensible authority? In other words, had the council represented that as clerk he had authority to contract?

Lord Justice Stephenson was emphatic:

'though there was nothing in [the council's] constitution prohibiting it from giving its clerk authority to contract on his behalf, there was no evidence that clerks to such authorities ever had that authority or that [this clerk] had authority to make contracts like this.'

Nor did he have any actual or implied authority.

The second question was whether, in any case, there was a binding contract including the fluctuations clause because of the form being signed and sealed on behalf of the council. The answer to this depended on the interpretation of what is now section 175 of the Local Government Act 1972, but was then section 266 of the 1933 Act. Here, the three appeal judges differed.

The minority view was expressed by Lord Justice Brandon who, having examined general contractual principles looked at the provision to section 266(2) which says:

'provided that a person entering into a contract with a local authority shall not be bound to inquire whether the standing orders ... have been complied with, and all contracts entered into by the local authority, *if otherwise valid*, shall have full force and effect'

even if standing orders have not been complied with. He read the italicised word as meaning 'valid apart from the failure to comply with the standing order', and on the facts he thought that the contract would have been otherwise valid.

This view was not shared by the other appeal judges. They took the view that the proviso did not validate a contract with a local authority deliberately and knowingly entered into by an agent in breach of the standing orders. Thus, the association was not protected by the proviso to validate the clerk's lack of authority. Since the clerk had no authority to contract on the council's behalf, there was no contract at all between the council and the association!

At a time when local authorities had to contract under seal, it had been held (in *A. R. Wright & Son Ltd* v. *Romford Corporation* (1956)) that a contract by a local authority not under seal could not be 'otherwise valid' under the proviso because it did not comply with the common law, so that the contract was invalid.

Lord Justice Stephenson applied the same rule.

'The only person who entered into or made [the] contract was [the clerk]. When he agreed to the association's unauthorised alteration of the contract authorised by his council, he and the association were *ad idem*, but his council and the association were no longer *ad idem*. There was no contract to be sealed. The sealing of the altered contract did not make the council and the association *ad idem* or make a contract where none was.'

It followed from this ruling that the contract was wholly void, with the consequence that the council could not recover any moneys paid under it.

However, while allowing the appeal, the court dismissed both the claim and counterclaim, leaving the association with a possible claim on *a quantum meruit* (as much as it is worth) if they could prove that they had been paid less than their work was worth.

This curious case is unlikely to be repeated. But it emphasises the need for care in contractual procedures, from both the employer's and contractor's point of view.

The need for precise provisions

Bill provisions drawn up by quantity surveyors are often imprecisely drafted and can give rise to problems if litigation ensues. This point is illustrated by *Convent Hospital Ltd* v. *Eberlin & Partners (1988)*, which discussed important issues about performance bonds and quantity surveying practice. It involved an employer's claim against architects who, it was alleged, had not properly performed the duties they owed to the employer because no performance bond had been entered into as required by the Contract Bills. The architects settled with the employer and, in turn, claimed against the insolvent contractors and its director for misrepresentation.

The bond requirement in the Bills was in a widely-used standard form:

'Provide for an approved bond ... in a sum amounting to 10% of the contract sum, for the due execution and completion of the works and for the payment of any damages, losses, costs, charges or expenses for which the contractors may become liable under the contract ... Note: The contractor must price this item and, if upon execution of the contract no bond ... is required by the employer, an appropriate deduction will be made from the contract sum in the final settlement of account.'

On the face of it, this provision is clear enough, and the case proceeded on the basis that this provision required that the approval of a bond must have been given on or before the execution of the contract and that the bond must have been in force at the time the contract was made.

His Honour Judge James Fox-Andrews QC did not have to decide whether that interpretation was correct, but seems to have been in doubt. He said:

'If it is the intention of the draughtsman that the provision should have such a meaning it is desirable that the wording makes this clearer than it does at present.'

One reading of the *Note* might suggest that the decision whether or not to require the bond would be left to the moment the contract was made.

'But, of course, in practice some time will elapse between an approach to underwriters or insurers for a bond and an indication by them to contractors or their brokers whether they were prepared to issue a bond and, if so, with what wording and at what premium.'

There is potential for dispute with Bill requirements for performance bonds unless they are carefully drawn. This is shown by an unreported decision of the same judge – *Arbiter Investments Ltd* v. *Wiltshier London Ltd* (1987), where the Bill requirement was that the bondsman be 'approved by the employer' and that the 'bond shall be presented to the employer for inspection and approval before commencement of the work.'

In that case Judge Fox-Andrews held that the employer's approval was subject to an implied term that the employer's approval should not be unreasonably withheld. He also ruled that the particular wording contemplated that the employer's inspection and approval of the bond could take place after the contract was made but before the commencement of the works.

In the *Convent Hospital* case, the judge thought that the wording of the *Note* to the Bill requirement was unsatisfactory and ambiguous. It referred to an appropriate deduction [being] 'made from the contract sum in the final settlement of accounts' if the employer required no bond. Disputes could arise at final account stage unless the deduction was merely that of the figure at which the item was priced. In practice, if the employer's quantity surveyor takes the view that the contractor has under-priced this item – which is often the case – he will seek a greater deduction from the priced item. This alone suggested that the wording of the *Note* was unsatisfactory.

In fact, the contractors had put in a low figure of £250 for a bond item and, since the contract sum was £1.3 million, they took a double risk in doing so. First, any quantity surveyor with experience of performance bonds was likely to identify the item as one to be challenged when considering the priced Bills before letting the contract. Second, the omission of the item would only be of limited attraction to the employer as the saving would be so small if, as so often happens, the employer found the lowest tender price was beyond his budget and a bill of reductions was negotiated.

Many contractors operate on the basis that in pricing the Bills their quantity surveyor seeks to identify items which may be omitted and puts in a lesser price than the internal price. The difference between the net cost to the contractor and the price in the Bill is then transferred to, and built into, other rates. Ideally, from the contractor's point of view, enhanced rates are entered for items where it is likely that the quantities will be increased.

Judge Fox-Andrews was firmly of the view that contractors cannot be criticised for operating in this way:

'The representations they were making were in no way false. They took the risk that financial loss would result to them if an item was priced below cost and the item was not omitted. Part of the skill required of the contractors was that enhanced rates for other items should not be too easily identified and challenged.'

In the event, no bond was entered into, and hence the employer's claim against the architects. When questioned about the low figure for the bond, the contractors' representative stated that they had special arrangements with an insurance company. That statement apparently persuaded the employer's quantity surveyor not to investigate the value of the item any further.

The court accepted expert evidence to the effect that the architects had an overall responsibility for the Bills of Quantities. It was essential that they should consider the Bill provisions to satisfy themselves that they were in the form that they required. The job architect, in fact, does not appear to have turned his mind to the question of the bond and so, although the judge found the contractors had made false representations about the bond, these had no effect on the architects who were not induced to act in any particular way as a result. Because of this, their claim against the insolvent contractor's director for misrepresentation failed.

Reliance on a seal

The practical reason for entering into a contract under seal is to ensure the benefit of the 12-year limitation period so that any claim will not be statute-barred until that time has elapsed. Whether the contract was made under seal and, if not, whether the employer was estopped from denying that it was so made was a point at issue in *Whittal Builders Co. Ltd* v. *Chester-le-Street District Council* (1988), Whittal's claim for damages related to alleged breaches of contract between 1972 and 1974, although the first intimation of any claim appeared in a letter from Whittal dated 28 May 1980. The writ was not issued until 5 January 1983 and so unless the contract was under seal the claim was barred.

The contract, made in 1972 in JCT 63 terms, was for the modernisation of 90 houses. The parties had contemplated that the contract would be sealed.

The articles of agreement sent by the council to Whittal had wafer seals attached. Whittal signed and sealed both copies but, through inadvertence, the council omitted to do so.

The chairman and clerk to the council signed the articles but the document was not sealed with the council's common seal, although the council had earlier resolved that it should be sealed. His Honour Judge James Fox-Andrews QC found that while a valid and binding agreement was reached, it was not one under seal. He said:

> 'It would be a legal fiction to hold that a document which palpably had not been sealed should nevertheless be sealed. Sometimes it is necessary to create a legal fiction.
>
> A court, however, should be slow to do so. I do not consider it proper to countenance that fiction here.'

But Whittal had a second string to its bow. It contended that the council was prevented from denying that the agreement was under seal on the basis of estoppel. Its case was that the council had represented that the agreement would be under seal, that in reliance on this Whittal sealed the contract and that it did not commence proceedings within the six-year limitation period which it would otherwise have done. The judge found that the council had represented that the contract would be under seal by putting it forward in a form appropriate for sealing. He decided that, on the strength of the council's representation, Whittal sealed the contract itself.

But that was not sufficient. The vital question was whether Whittal had altered its position so that it suffered detriment in reliance on that representation. The judge ruled:

> 'The test of detriment is whether it appears unjust or inequitable that the defendants should not be allowed to resile from the representation, having regard to what Whittal has done or has refrained from doing in reliance on that representation.'

The evidence did not satisfy the learned judge that but for the fact that Whittal believed the contract was under seal Whittal would have started proceedings within the six-year period. He was not satisfied that Whittal had received any advice on limitation matters or applied itself in any way to the significance of sealing until it consulted solicitors in 1980. By then any claim under a simple contract would have been statute-barred. The case based on *estoppel in pais* therefore failed.

However, this did not dispose of the issue because at the time when they entered into the contract both parties assumed as a basis of the transaction that the contract would be under seal. This gave rise to an estoppel by convention so that each party was now prevented from questioning the truth of the assumption. Whittal was therefore successful on this point, but the court had to consider the circumstances of the contract and the terms in which it was made, since Whittal's substantive claim was for breach of the obligation to give possession of the site.

At the time the parties were negotiating the contract, the building industry was on a lengthy strike. It was therefore in the contemplation of the parties that the strike was likely to be continuing not only when the contract was made but also on the appendix date for possession.

Thus, although the contract provided for the council to give possession on a stated date, on its true interpretation the contract meant that possession was to be given on the stipulated date or the date when the strike ended, whichever was earlier.

Clause 21 of the contract provided for the contractor to be given 'possession of the site' on the due date. 'After some hesitation', Judge Fox-Andrews held that this meant possession of all 90 dwellings as well as the area for the site office, despite the fact that in a modernisation contract it would be unusual for the contractor to be given possession of all the houses at the same time.

Possession had been given piecemeal and the council had been in breach of contract. *Prima facie* Whittal was entitled to damages, as explained by Justice Vinelott in *London Borough of Merton* v. *Stanley Hugh Leach Ltd* (1985).

As Judge Fox-Andrews put it:

'Although the JCT contract does not expressly provide for the situation where the employer fails to give possession on due date, I have no doubt that a claim for damages may be pursued by a contractor who suffers loss and damage as the result of such a breach.'

But in the event, the contractor's victory on these two points was a pyrrhic one because, so the court found, it was aware from September 1972 onwards that it had a claim for damages arising out of the delayed possession and on the facts in March 1977 the parties had entered into an agreement in full and final settlement. The council had sent its cheque and the judge was satisfied on the evidence that the payment was in respect of all claims and cross-claims. Alternatively, said the judge, by its silence up to March 1977 Whittal had waived any claim arising out of late possession or was itself estopped from raising such a claim thereafter.

Chapter 2

Interpretation and Implied Terms

The rules of interpretation

Once a contract has been reduced to writing, 'verbal evidence is not allowed to be given ... so as to add to or subtract from, or in any manner to vary or qualify the written contract': *Goss* v. *Nugent* (1833).

This basic rule of interpretation is called 'the parol evidence rule' and covers not only oral evidence, but also other evidence outside the document: drafts, pre-contract letters and so on. Equally, it prevents evidence being given of the preliminary negotiations between the parties.

Like so many other rules, it is subject to exceptions. The rule does not apply, for example, where misrepresentation is alleged and, indeed, where one party claims that there is a collateral contract or warranty. It remains, however, a basic rule of interpretation of written or printed contracts and reference may usefully be made to Chapter 9 of *Making Commercial Contracts* by John Parris for a discussion of what he calls 'one of the strangest but well established principles of construing a contractual document.'

Construction contracts are generally in a printed standard form. Various optional clauses are deleted and there may be typewritten or manuscript amendments. Logically, therefore, the parol evidence rule would preclude an arbitrator or the court from looking at what was deleted.

In fact, in *Mottram Consultants Ltd* v. *Bernard Sunley & Sons Ltd* (1974) – one of the 'set-off' cases – the House of Lords held that in interpreting a printed contract, one is entitled to look at the words that had been deleted 'as part of the surrounding circumstances in the light of which one must construe what they have chosen to leave in'.

'Surrounding circumstances' is an imprecise phrase which can be illustrated but hardly defined, as Lord Wilberforce pointed out in a shipping case.

'In a commercial contract it is certainly right that the court should know the commercial purpose of the contract and this in turn presupposes knowledge of the genesis of the transaction, the background, the context, the market in which the parties are operating',

and so on.

Extrinsic evidence will also be admitted to explain the written agreement, and in particular to show the meaning of individual words and phrases used by the parties. The starting point for finding the meaning of a word is the dictionary, but both courts and arbitrators must give effect to any special technical, trade or customary meaning which the parties intended the word to bear.

The contract may itself contain an interpretation clause, as does JCT 80, clause 1 of which sets out some definitions. In fact, most of those definitions are far from helpful, but that is not the point. Clause 1 of the ICE Conditions of Contract, 5th edition, is also a definition clause and seems to be of more practical use.

The main basic rule of interpretation is that the agreement must be read as a whole – a particular clause must be seen in context. In *Brodie* v. *Cardiff Corporation* (1919), Lord Atkinson put the point in this way:

'The contract must be construed as a whole, effect being given, so far as practicable, to each of its provisions.'

This is a point often overlooked by those without formal legal training who, in argument, will seize on a particular word or phrase out of context.

In the construction industry, the definition of the 'contract documents' is important; all well-drafted contracts will give a clear definition of them. Thus, JCT 80, clause 2.1, and ACA 2, 1984, article C, give a comprehensive definition, and in interpreting the contract it is to these documents we look.

The wording of the contract may also attempt to introduce interpretative rules of its own, and these may reverse the normal legal rules. So it is with JCT 80, clause 2.2 which has the effect of making the printed conditions prevail over typed or handwritten documents, which reverses the normal and logical rule.

The validity of this provision has been upheld time and time again, and despite the strictures of Lord Denning M R, as he then was, in *English Industrial Estates* v. *George Wimpey & Co. Ltd* (1973), the printed conditions will prevail over any typed provisions in the bills or specification.

At the very best, to quote Lord Justice Edmund Davies, the bills etc.

may be used 'not in the interpretation of the contract ... but in order to follow exactly what is going on ...' and presumably as part of the 'surrounding circumstances'.

There is another basic rule which is sometimes invoked and it is called the *contra proferentem* rule.

'If there is an ambiguity in a document which all the other methods of (interpretation) have failed to resolve so that there are two alternative meanings to certain words the court may construe the words against the party seeking to rely on them and give effect to the meaning more favourable to the other party'. D. Keating, *Building Contracts*, 4th edition, p. 34.

Unfortunately for the building team, the rule does not seem to apply to 'negotiated' standard form contracts, such as JCT and ICE contracts, where the document is prepared by representatives of actual and potential users: see *Tersons Ltd* v. *Stevenage Development Corporation* (1963).

Probably, however, the rule would apply where the employer makes *substantial* amendments to the printed text so that it ceases to be a 'negotiated document' and is put foward by him as his own. The rule was applied in the case of Liverpool Corporation's own form of building contract in the oft-quoted Court of Appeal ruling in *Peak* v. *McKinney* (1970).

The court may well disregard completely meaningless words, but the judicial task is to interpret the intentions of the parties and not to write a contract for them. In some famous cases, Lord Denning, when Master of the Rolls, came very close to doing just that. Apparent inconsistencies between clauses in a contract will be reconciled if it is possible to do so, otherwise the court will give effect to the clause which, in its view, expresses the true intention of the parties.

In setting out some of the rules used to interpret contracts I do not seek to justify them, though – by and large – they are logical enough. The rules are, however, very important for it is by them that what has been expressed will be interpreted.

Redefining the boundaries

The majority of contracts used in the construction industry comprise comprehensive written agreements, but in closely defined circumstances the courts will write in 'implied terms', usually to make the agreement commercially effective. Terms will not be implied merely to get the parties out of difficulties, nor will the courts rewrite the contract which the parties have made.

The difficulties facing someone invoking an implied term are vividly illustrated by *Barratt Southampton Ltd* v. *Fairclough Building Ltd* (1988), where the parties exchanged two housing estates with an equalising payment. Some of the roads and drains on the estate received by Barratt were alleged to be defective. Under clause 4.8 of the specially drafted exchange agreement Fairclough warranted that the roads were 'constructed in their entirety to base course to the appropriate lines and levels as approved and inspected by' the local authority.

Barratt argued that this constituted a warranty that the roads and drains had been so constructed in accordance with good construction practice and to a standard fit for adoption by the council. Alternatively, it said that terms to the same effect were to be implied as a matter of 'practice in the building industry' and/or to give business efficacy to the express terms agreed between the parties. His Honour Judge Peter Bowsher QC rejected Fairclough's argument that the express words of the warranty could have no meaning other than their literal one. English words are rich in meaning and, even in commercial documents it is essential to determine the meaning of words in their context.

There was no evidence of 'practice in the building industry'. Fairclough submitted that the agreement was effectively one to convey property, while Barratt said that it was a kind of half-way house. The basic rule is that in the absence of an express contract the seller of real estate is not liable to the purchaser for defects in the property which make it dangerous or unfit for occupation, even if he knew of the defects or was responsible for them. Barratt's argument was that either an express or implied contract was in existence which negated this rule.

Clause 4.8 of the exchange agreement began by saying that both estates were 'sold in their present state and condition', and in the event the judge ruled that clause 4.8 must be interpreted literally. The agreement was one to sell real property and, while it was essential that the roads and drains were up to the standards required by the local authority, it was not necessary either for the exchange project or the purposes of the agreement that Fairclough would be liable for any failure to meet those standards. It was not necessary to imply a term to make the agreement workable. The judge said:

'If an express warranty as to standards of the roads had been insisted upon, that would be a matter which might have been expected to influence the amount paid by way of the balancing charge.'

Barratt's claim therefore failed; the contract risk was its own. It was not necessary to imply a term or to give a gloss to the plain words of clause 4.8.

In reaching this conclusion, Judge Bowsher examined the authorities restricting the implication of terms in contracts. He usefully summarised the three main circumstances in which terms have been implied.

- *By implication of law*
  In cases such as sales of goods and bills of exchange the courts imposed implied terms on the parties by way of judicial legislation. These have been consolidated by statute, and now have the status of standard terms of contract upon which everyone contracts unless express contrary terms are agreed. The judge observed:

  'There is little to be learnt from them when considering what terms should be implied in a modern commercial contract falling outside the ambit of those old cases or the consolidating statutes'.

- *Necessity*
  In some cases, where there is an apparently complete bargain, the courts are willing to add a term on the ground that without it the contract will not work. Any term of this class has to be reasonable but that is not sufficient. The cases show that the implication of this class of term is based on the parties' presumed intention. It is called *The Moorcock* doctrine, and derives from something said by Lord Justice Bowen in the nineteenth-century case of that name. In modern commercial practice there is little ground for implying terms on the basis of necessity. On the facts the exchange agreement was commercially effective as it stood.

  But in ordinary building contracts terms of co-operation by the employer and against hindrance or prevention by him may be implied on this basis: *London Borough of Merton* v. *Leach Ltd* (1985). Under building contracts there is a general obligation as to co-operation because otherwise the contract's purpose would be frustrated. In sophisticated forms of contract, this obligation may be excluded, as a reading of the *Leach* case shows. Indeed, the Privy Council have indicated that where there is a detailed contract there is little room for the implication of terms: *Tai Hing Cotton Mill* v. *Liu Chong Hing Bank* (1986).

  Keating's *Building Contracts*, 4th edition, p. 37, makes the point that in modern building contracts: 'there may be little room for the implication of any terms for if the parties have dealt expressly with them in the contract no terms dealing with the same subject can be implied.'

- *The officious bystander*

  The third case in which a term might be implied, though in many instances it does not differ from terms implied from necessity, is based on Lord Justice Mackinnon's statement in *Shirlaw* v. *Southern Foundries Ltd* (1939) to the effect that a term may be implied if it:

  'is something so obvious that it goes without saying, so that if, while the parties were making their bargain, an officious bystander were to suggest some express provision for it, they would testily suppress him with a common "Oh, of course"'. Such cases are rare in practice.

The difficulty of relying on an implied term is well shown by *Barratt* v. *Fairclough*, and in most circumstances the argument for an implied term is one of last resort. It is now abundantly clear that judicial legislation is unlikely to have much effect in modern commercial contracts.

The current position is further indicated by another decision of Judge Bowsher in *University of Glasgow* v. *Whitfield* (1988) where the contract was in JCT 63 form. He found that there was no implied term that the contractor should warn the employer about possible defects in design by the architect. This followed from *Tai Hing Cotton Mill* v. *Liu Chong Hing Bank* (1986) which was primarily concerned with the point that where a contract expressly deals with the subject matter of a claim there is no room for a parallel duty in tort. However, he also referred with approval to the Court of Appeal decision in *Lynch* v. *Thorne* (1956). Lord Evershed MR summarised the position in that case as follows:

'Where there is a written contract expressly setting forth the bargain between the parties it is, as a general rule, also well established that you only imply terms under the necessity of some compulsion.'

The boundaries have been redefined.

It goes without saying

One of the commonest problems with modernisation contracts is that of properties being handed over in groups so that the contractor can progress the work in an orderly manner. Difficulties arise in both legal and practical terms, especially when standard form contracts are used.

The question was one of the points at issue in *Whittal Builders Co. Ltd v. Chester-le-Street District Council* (1985), where the contract was in JCT 63 form. It was a fixed price contract for the modernisation of 108 houses. At pre-contract stage it was agreed that the work should be completed in 18 months provided Whittal could have at least 18 houses at any one time and, so the contractors alleged, in pairs. All the houses were terraced. The major part of the work to be done was the building of a rear extension which, in all but six cases, was a single building extending out to the rear of two houses, divided by an internal party wall.

The possession clause – clause 21 – was unamended and provided that 'on the date for possession stated in the Appendix ... possession of the site shall be given to the contractor ...' The Appendix entry gave the date for possession as 15 October 1973 and the date for completion as 14 April 1975, i.e. 18 months later. However, after the date for completion the words 'provided that 18 dwellings be available for the contractor to work in at any given time'.

The Council argued that those words related only to the contractor's obligation and that their only effect was that if the employer did not give possession of at least 18 houses at a time, the contractor's time for completion was extended. There was no duty on them, they said, to give possession at that or any other particular rate.

That argument found no favour with the judge, Mr Recorder Percival. He said:

'The view which I take is that the words relating to 18 houses apply both to the obligations of the employer and the contractor. The plain meaning of the words in clause 21 read together with the relevant words in the Appendix is that the employer is obliged to give possession of at least 18 houses at any given time starting on October 15 1973 (in effect his absolute obligation is cut down to that extent) and the contractor's obligation to complete by April 14 1975 is made conditional on the employer giving him not less than 18 houses at any one time.'

The judge was also firmly of the view that the employer was bound to give possession of the houses in pairs, and he reached that conclusion by two different routes.

- As a matter of contract interpretation that requirement was to be spelled out from the contract documents as a whole.

'Where there is a clear obligation on the contractor to build the extension as one single building on a site extending over and forming part of each of two separate properties, that necessarily … imports an obligation on the part of the employer to give possession of the whole of such a site at one time.'

Put simply, the employer's duty was to give possession of the houses forming a pair at the same time.

- Alternatively, ruled the judge, a term to the same effect should be implied that, except in the case of six individual single houses, possession was to be given in pairs.

The learned judge referred to the well-known case of *Trollope & Colls Ltd* v. *North West Metropolitan Regional Hospital Board* (1973), which was concerned with implied terms in a building contract. In his opinion all the tests specified in that case were met in full.

'The matter is to me so obvious that I have no doubt that tests such as: "We did not trouble to say that, it is too clear"; "It goes without saying"; and "It must have been a term which went without saying" are as obviously and fully met in this case as it is possible to imagine.'

So he found in favour of the contractors on this point.

Whittal Builders Co. Ltd v. *Chester-le-Street District Council* suggests an answer to another common problem, namely what happens when all the technicalities regarding making a contract under seal are not observed. The point was of great importance because if the contract was under seal the contractor's claim was not statute-barred as it had an extended limitation period of 12 years under section 8 of the Limitation Act 1980. Through an oversight, the company failed to put its impressed seal on the red wafer stuck to the formal contract received from the Council, although it was stamped with the Council's seal. The contract was also unstamped, as then required. The Council said that someone wishing to make a contract under seal must 'get it right to the letter'.

Again, the judge did not agree. He quoted the words of Mr Justice Danckwerts in *Stromdale & Ball* v. *Burden* (1952):

'Meticulous persons executing a deed may still place their finger on the wax seal or wafer on the document, but at the present day if a person signs a document bearing a wax or wafer or other indication of a seal, with the intention of executing the document as a deed, that is sufficient adoption or recognition of the seal to amount to due execution as a deed.'

In consequence the judge ruled:

- The omission of the company's seal did not affect the validity of the contract.
- No different standard should be applied to acts done by limited companies or individuals.
- The contract was one under seal and Whittal's claim was not statute-barred.

He also took the view that in such a case:

'all that was material was whether the formalities were complied with by the party to be charged [ie the Council] and that so far as the [Council] was concerned no question or doubt whatever could be raised either as to signing, sealing or delivery'.

Implied terms in JCT contracts

The terms to be implied into a JCT contract were one of the many points at issue in *London Borough of Merton* v. *Stanley Hugh Leach Ltd* (1985). It is by means of implied terms that even a lengthy formal contract may be made workable, and the parties' presumed but unexpressed intentions may be given effect, subject to the limitations discussed in earlier essays.

The contractor alleged that two main terms should be written into the contract:

- That the employer would not hinder or prevent it from carrying out its contractual obligations, or from executing the works in a regular and orderly manner.
- That the employer would take all steps reasonably necessary to enable it to discharge its obligations under the contract, and to execute the works in a regular and orderly manner.

These are the terms usually pleaded in any well-drafted set of arbitration documents, and neither the arbitrator nor the judge had any difficulty as regards the first term. There was no express term in the contract excluding this implied term which was necessary to give business effectiveness to the contract.

Indeed, as Mr Justice Vinelott rightly said, 'It is difficult to conceive of a case in which this duty could be wholly excluded'. A duty of co-operation between the parties is essential to make the contract work. It is not the less necessary because the JCT contracts set out a detailed and comprehensive skein of rights and obligations, some of which do enable the employer to interfere with the contractor's progress, e.g., clause 29 of JCT 80 which entitles the employer to introduce his own directly-employed licensees on site.

There are many other actions of the employer or his architect which can hinder the contractor in his progress, and result in delay to completion, or the contractor's involvement in loss and expense. For example, an architect might instruct the contractor to deposit spoil on part of the site, and then fail to instruct its removal at such a time as would allow the contractor to use that part of the site at the time he planned to do so.

This implied term clearly fills in the gaps in JCT contracts – and other standard form contracts – and enables the contractor to recover for any damage which he suffers as a result of its breach. Where the JCT contract permits an employer to interfere with or hinder the contractor's progress so as to cause him loss or expense, then it provides for him to be reimbursed under the contract machinery. Other acts of hindrance or prevention must, therefore, be covered by an implied term.

Both judge and arbitrator were also of the view that the second term should be implied. The parties are under a duty to co-operate whenever it is reasonably necessary, to enable the other to perform his contractual obligations. However, as the judge rightly pointed out, 'the requirement of "good faith" in systems derived from Roman law has not been imported into English law', and the duty is a fairly limited one. In effect, the second term was the positive aspect of the first implied term.

Again, while the JCT contract imposes express duties of co-operation – such as the architect's duty to issue instructions at the right time – and provides remedies for breach of those duties, it is a contract drafted for general use and cannot be absolutely exhaustive in its provisions. The implication of the term was necessary to give business effectiveness to the contract.

The contractor contended for other terms to be implied. One, in a form modified by the arbitrator, was that the architect would provide the contractor with correct information concerning the works.

The employer argued that the express terms of the contract made this term otiose, but both judge and arbitrator held that the term was to be implied. It must, said the judge, have been in the parties' contemplation that the architect would act with reasonable diligence and use reasonable care and skill in providing the contractor with information, e.g. under clause 5.4 which imposes on him an obligation to provide the contractor with drawings and details, as and when necessary.

The judge emphasised – countrary to the view espoused by many architects – that:

> 'the contract does not impose a duty on a contractor to check the drawings to see if there are discrepancies or divergencies, and discrepancies or divergencies may come to light at a time when, it is too late for the contractor to call for an instruction, or draw in further drawings and details under clause 5.4.'

These words should be emblazoned in every architect's office!

The fact that a contractor has a remedy when discrepancies or divergencies are discovered in good time, does not affect the principle for, as Mr Justice Vinelott remarked, this:

> 'would not be regarded by a reasonable man familiar with the operation of the contract, as absolving the building owner from liability for the architect's failure to use reasonable care and skill in supplying drawings and other information, in a case where that machinery does not afford the contractor an adequate remedy'.

The very point about implied terms is that they are intended to fill the gaps in the contract. Terms will not be implied which are repetitive of, or repugnant to, the express terms of the contract and, to quote Lord Diplock in *Photoproductions Ltd* v. *Securicor Ltd* (1980), the:

> 'parties are free to agree to agree whatever exclusions or modifications of all kinds of obligations as they please, within the limits that the agreement must retain the legal characteristics of a contract'.

Despite *London Borough of Merton* v. *Stanley Hugh Leach Ltd*, the dispute about the terms to be implied in JCT contracts is unlikely to be stilled. The importance of this aspect of the case is that it is a direct ruling on implication of terms in a JCT contract, admittedly in JCT 63 form. The principles are of equal application to the other JCT forms.

Standards of quality

It has long been settled law that terms as to fitness for purpose would be implied into a contract for the sale of goods. The position has now been codified in the Sale of Goods Act 1979. Buildings are not, of course, goods and it is only in comparatively recent years that a similar term has been implied in respect of houses and other structures.

In some cases statute affects the position. So, dwellings are subject to the provisions of the Defective Premises Act 1972, and the duty imposed by the Act is additional to any other duty which a contractor or developer may owe apart from the Act. Section 1(1) provides:

> '*Any person* taking on work for or in connection with the provision of a dwelling ... owes a duty to see that the work he takes on is done in a workmanlike manner, with proper materials ... and *so as to be fit for the purpose required* ...'

Clearly, this provision imposes far more extensive obligations on builders and developers and others – including architects – than are contained in the standard form construction contracts.

Unfortunately there has been no case law development, though there have been judicial observations on the possible effects of the Act.

The ambiguous position of contractors working under JCT 80 is discussed by John Parris in *The Standard Form of Building Contract: JCT 80* (2nd edition) where he refers to the provisions of section 1(2) of the Act, which says:

> 'A person who takes on any such work for another on terms that he is to do it in accordance with instructions given by or on behalf of that other shall, to the extent that he does it properly in accordance with those instructions, be treated ... as discharging the [statutory duty] except where he owes a duty to that other to warn him of any defects in the instructions and failed to discharge that duty'.

Since JCT 80 requires the contractor to comply with 'all instructions issued to him by the architect' under the contract, he concludes that the contractor will be able to rely on the defence:

> 'provided, of course, that he has complied with the architect's instructions, has given appropriate warning of any defects in those instructions, and provided a reasonably competent contractor would have perceived such defects'.

This seems a sound argument, but of course we shall not know the answer until some brave litigant faces the issue squarely. Quite apart from the Act, the contractor may owe duties at common law, though the standard form contracts all cut down on what would otherwise be his obligations.

The development of the law as admirably summarised by His Honour Judge John Newey QC in *Basildon District Council* v. *J. E. Lesser Properties Ltd* (1987), a case involving system builders, who were working under a JCT 63 contract with amendments.

On the facts and contract documentation the learned judge ruled, *inter alia*, that it was an implied term of the contract that the buildings as designed by the contractors should be fit for human habitation on completion. In reality, the design work was their exclusive responsibility. They had prepared the drawings and the responsibility continued even after they had become incorporated into the contract. Moreover, the employer had relied on the contractor's expertise as a system builder to produce habitable dwellings. Alternatively, ruled the judge, there should be implied a lesser term to the effect that the contractor would design with the skill and care to be expected of a system builder.

In reaching this conclusion, His Honour considered the development of the law in light of all the major cases. For example, in *Miller* v. *Cannon Hill Estates Ltd* (1931), the defendant had agreed to build a house for sale to the plaintiff. It had contracted that it would use the best workmanship and materials. Mr Justice Swift held that there was also an implied warranty that the house would be fit for human habitation on completion.

'[The] whole object, as both parties knew, is that there shall be erected a house in which the intended purchaser shall come to live. It is the very nature and essence of the transaction ... that he will have a house put up there which is fit for him to come into as a dwelling house. It is plain that in those circumstances there is an implication of law that the house will be reasonably fit for the purpose for which it is required – that is for human dwelling'.

After considering the various other authorities which developed the law, including the famous case of *Independent Broadcasting Authority* v. *EMI Electronics Ltd and BICC Construction Ltd* (1980), a decision of the House of Lords, he had no hesitation in upholding the implied terms contended for.

He quoted with approval the basis of the decision as stated by Lord Justice Roskill in the Court of Appeal:

'We see no reason ... for not importing an obligation as to reasonable fitness for purpose into those contracts or for importing a different obligation in relation to design from the obligation which plainly exists in relation to materials'.

Other recent cases have developed the law about design and build contracts, though of course where the JCT Standard Form of Contract with Contractor's Design is used, the contractor's obligation is more limited.

Clause 2.51 states 'insofar as the design of the Works is comprised in the Contractor's Proposals ...' he has 'the like liability ... as would an architect ... or other appropriate professional designer holding himself out as competent to take on work for such design'. In other words, liability will depend on proof of negligence and the contractor is not guaranteeing the result.

How many employers using that form, I wonder, realise the legal position? They would do better to contract at common law!

A problem with a sub-contract

Older cases – as well as new ones – are of interest because once the courts have handed down a decision, the industry is on notice as to its effect. Mistake of law is no defence, and it is foolhardy for construction professionals to ignore clear principles of law as laid down in judicial decisions. Those cases also illustrate the operation of legal rules in a factual situation.

The pre-War decision of the Court of Appeal in *Chandler Brothers Ltd v. Boswell* (1936), is a case in point. It lays down a general rule of much practical interest and importance about the terms which will be implied into a building contract. This rule is often ignored in practice by employers and contractors alike – and sometimes this ignorance extends to professional advisers.

It was a civil engineering case. Boswell contracted with Caernarvon County Council to divert a road. The contract included tunnelling, which was sub-contracted to Chandler. Soon after starting work, Chandler went into receivership. The tunnelling work was approaching completion and, as he was bound to do, the receiver carried on with the sub-contract. For financial reasons, the receiver was unable to proceed as quickly as the sub-contract terms required.

Boswell's contract with the council contained a special clause covering sub-contractors. It read:

'If the engineer shall at any time be dissatisfied with any sub-
contractor or other person similarly employed or engaged by the
contractor upon, or in connection with the works, the contractor
shall, if required in writing by the engineer to do so, forthwith
remove him, or them, and put an end to his or their employment or
engagement.'

There was no corresponding provision in the sub-contract. Boswell
acted upon an engineer's instruction to terminate. In doing so, he was
held by the Court of Appeal to be in breach of his sub-contract with
Chandler.

The matter was very fully argued before the Court, largely on the
basis that there was a term to be implied in the sub-contract giving the
contractor power, as between him and the sub-contractor, to put an
end to the sub-contract should the engineer give notice under the
main contract clause.

Lord Justice Greer dealt with this argument briefly:

'It is not possible to say that the event that happened in this case was
not actually present in the minds of the parties at the time of making
the contract, because it seems to me quite clear that they had before
them, and used for the purposes of the sub-contract, clauses by
transposition which were in the original contract, and they omitted
to include [the notice clause], showing quite clearly that they had in
their contemplation ... all the clauses of the head contract, and
intentionally refrained from making any reference to what was to
happen if [the notice clause] were put into operation by the
engineer.'

The essential point is, therefore, that main and sub-contract should
tie in together exactly – and if there is a determination of forfeiture
clause in the main contract, the sub-contract should parallel its
provisions. This is, of course, done by all the standard form contracts,
whether of the JCT family or otherwise. But it is not done in the case
of home-drawn versions. The problem also arises where, for example,
there is no formal sub-contract, but merely an order from a main
contractor to a sub-contractor. This is far more common than is
generally supposed.

The result in *Boswell's* case was predictable, although a further
argument was advanced before the Court of Appeal. This was that the
notice clause was incorporated into the sub-contract by reason of a
sub-contract recital that the sub-contractor had agreed to carry out
the work in accordance with the terms of the main contract. That

argument was also rejected. The recital only meant that Chandler was to provide work of the quality and with the speed of progress which was stipulated for in the main contract itself. It did not incorporate the forfeiture clause.

Another interesting point was discussed, namely that the slowing-down of the sub-contractor's work through lack of money, was itself repudiatory conduct amounting to breach such as would entitle the main contractor putting an end to the contract. Lord Justice Greer dismissed that argument tersely:

> 'It is not right for everybody to do with speed that which the contract requires the contractor to do with speed. It is not every small breach of contract which justifies putting an end to the contract at common law.'

In support of this he relied on the House of Lords decision in *Mersey Steel & Iron Co. Ltd* v. *Naylor, Benzon & Co.* (1884), where in a supply contract buyers postponed payment under erroneous legal advice. It was there held, among other things, that this did not show any intention on the part of the buyers to repudiate the contract, or to release the sellers from further performance of their side of the bargain.

In the circumstances of the case, Boswell was in breach of his sub-contract on all counts. The effect of Boswell giving wrongful notice was to give Chandler the right to sue for breach of contract or on a *quantum meruit* ('as much as it is worth') basis. He was not obliged to wait until the date for completion arrived, and then sue on a breach of contract at that stage. He could do so, but the option was his.

In cases like this, therefore, terms will not be implied into a subsidiary document, such as a sub-contract to provide for any result ensuing from main contract provisions where the parties have sat down with the main contract documents before them. The sub-contractor's rights and duties are not governed by the terms of the main contract, unless they are incorporated into the sub-contract. The contractual relationship between main and sub-contractor is independent of that existing between main contractor and employer. It seems, too, that it is not good enough to state that the sub-contractor is bound by all the terms of the main contract, and it is absolutely essential that both main and sub-contractors ensure that the sub-contract documentation is compatible with the main contract requirements. Failure to take this precaution is a recipe for disaster.

'Unfair terms'

Whenever contractors gather together there is talk of 'unfair' terms in contracts. What is usually meant by this is that there is a clause in a particular contract which imposes liability unfairly – or so the complainant thinks.

The complaint may be, for example, that there is a heavy potential liability to liquidated damages, or that the contract period is too tight. More commonly these days – particularly in some local authority contracts – the wording of a standard 'direct loss and/or expense' clause such as JCT 80, clause 23, has been amended to the effect that 'direct loss and/or expense' is specially defined so as to exclude the financial burden or 'interest' element which would otherwise be included.

These are not 'unfair terms' in the legal sense. Indeed, those two words are not a term of art, and the Unfair Contract Terms Act 1977 does not deal with 'unfair terms'; it covers unfair or unreasonable exemption clauses, and these are a very different thing.

Section 2(1) of the Act outlaws completely clauses or notices which try and exclude or restrict liability in negligence for personal injury or death. Other sections make totally ineffective the exclusion or restriction of certain implied undertakings in contracts of sale or hire purchase.

Other contract terms are subjected to a test of 'reasonableness'. These include provisions attempting to limit liability for loss or damage arising from negligence other than personal injury or death and are covered by section 3. I err in good company if I find the provision complicated, since the leading practitioner's book on contract law (Cheshire Fifoot & Furmston's *Law of Contract*, 11th edition, 1986, p. 173) says that section 3 'contains a complex set of provisions, which are far from easy to understand or interpret'! Section 3 applies only where someone is dealing with a 'consumer' *or on his own written standard terms of business*. It provides that he:

> 'cannot by reference to any contract term – (a) when himself in breach of contract, exclude or restrict any liability of his in respect of the breach; or (b) claim to be entitled – (i) to render a contractual performance substantially diferent from that which was reasonably expected of him; or (ii) in respect of the whole or part of his contractual obligations, to render no performance at all'

unless the term meets the reasonableness test. Furmston's view is that

(b) catches not only ingeniously drafted exemption clauses but clauses which are not true exemption clauses at all.

The example given is that of a machine tool supplier who provides in his standard conditions that payment terms are 25% with order and 75% on delivery but that he is not obliged to start manufacture until he receives the first payment. As the author says, he might well pass the test with flying colours 'but it is not a good argument for putting hurdles on a motorway that most cars will drive through them'!

The industry-wide negotiated contracts such as JCT and ICE forms are not, in any case, the employer's 'written standard terms of business'. Specially-drafted in-house forms are – and substantial amendments to the standard forms may well bring them into the same category. There have been very few cases on the application of the reasonableness test, on which the Act sets out some very broad guidelines. The main point (section 11(1)) is that the question of reasonableness is to be decided:

> 'having regard to the circumstances which were, or ought reasona-
> bly to have been, known to or in the contemplation of the parties
> *when the contract was made.*'

The leading and indeed only construction industry case is *ReesHough Ltd* v. *Redland Reinforced Plastics Ltd* (1984), which emphasises how much discretion the judge has in applying the test. Redland's standard conditions of sale at issue accepted only limited liability for defective goods and went on to exclude all liability:

> 'contractual or tortious, in respect of loss or damage suffered ... as
> a result of any defect in the goods ... or as a result of any warranty,
> representation, conduct or negligence by'

Redland or its agents, and so forth.

His Honour Judge John Newey QC ruled that Redland had failed to prove that the exclusions were reasonable. Those concerned with similar problems ought to read his judgment in detail. The weightiest argument in favour of upholding the term was that Rees-Hough was able to look after itself and that Redland had never sought to rely on the terms in past dealings with them. But Redland failed to prove reasonableness and, as the judge remarked, 'the balance of considerations is strongly against the terms being reasonable.'

The position is probably different if standard terms put forward by one party are modified by negotiation – which is a common practice in the construction industry.

In fact, it is easier to justify reasonableness in relation to a limitation of liability than it is where there is a total exclusion of liability. This follows from the decision of Mr Justice Staughton in *Stag Line Ltd* v. *Tyne Shiprepair Group Ltd* (1984), although the contractor's terms were held to be ineffective because they imposed an unreasonable condition.

They set out to deprive the employer of *any* remedy unless the ship was returned to the contractor's yard for the defect to be remedied. One of the guidelines in the Act is a condition like this with which it may not be practicable to comply. One can think of similar situations in the construction industry in the standard terms of suppliers of equipment. However, the judge emphasised that:

> 'the courts should be slow to find clauses in commercial contracts made between parties of equal bargaining power to be unfair or unreasonable.'

It is curious that there is so little case law since the Act has been in operation for so many years. It applies to all contracts made after 11 February 1978. That probably says something for the good sense of the business community, if not for the ingenuity of the draftsman.

Ineffective exclusion clauses

Exclusion clauses are always interpreted strictly by the courts. If there is any doubt as to the meaning and scope of an exemption clause, the ambiguity will be resolved against the party who is relying on its wording to escape liability. It is for him to prove that the words used clearly cover the disputed situation. The position is well-illustrated by the decision of the Court of Appeal in *Ackerman* v. *Protim Services Ltd* (1988) where a specialist in the treatment of dry rot sought to deny liability under the terms of its 20-year guarantee.

Protim carried out work on the plaintiff's premises in north London during 1974. In 1982 there was a fresh outbreak of dry rot in the same area that had been treated eight years before. The original outbreak of dry rot was in a first-floor room. The trial judge found that one of the two areas affected was in the front wall of the building where the sill of a balcony – which ran the full length of the room – joined the floorboards. Underneath them was a timber bressumer, or wall-plate, which was partly embedded in the brickwork of the front wall.

The trial judge also found as facts that the treatment applied to that area in 1974 was bad because the specialist company knew, or ought to have known, that part of the bressumer which had been affected by dry rot had not been removed before treatment. Protim's operative sprayed over the affected area, which in all probability was visible to him. The dormant infestation was reactivated by the damp inner wall face in 1982.

The plaintiff's claims at common law were statute-barred and so he relied on the guarantee. This was expressed not to apply where:

> 'timbers or walls of the property ... suffer a recurrence of fungal attack (i) as a result of failure to keep the property in a dry and weather-proof condition and in good and proper state of maintenance ... (iii) as a result of a failure to carry out any recommendation given by the company in writing which is the responsibility of the client'.

The guarantee ran with the property for the benefit of subsequent owners.

The plaintiff sued under the guarantee for the cost of the remedial work. Protim refused to accept liability and relied on the terms of clause B(iii) of the guarantee. Its written recommendations included, by virtue of the contract documents, the removal of all timber built in walls and of all infested timber. The trial judge gave judgment for the plaintiff for £200; Protim was held liable to the 20-year guarantee on its appeal. Lord Justice Kerr described Protim's defence as 'discreditable' as it would render the guarantee largely worthless. The Court of Appeal held that, taking the guarantee as a whole, the vital words were 'as a result of failure'. This referred to some act or omission of the owner or occupier. It implied that something happened which ought to have been prevented by the owner or occupier. It was not enough to show that there was dampness sufficient to trigger the reinfestation in combination with the other circumstances. There was no evidence of 'failure' by any owner or occupier.

The proviso to clause B(iii) did not apply because the recurrence of the problem was not the result of any failure of the plaintiff. On the facts found by the judge, Protim's operatives were clearly wrong and at fault in proceeding with the treatment and allowing the area to be resealed when they knew or ought to have known that it was improperly treated and still infected. Protim's appeal was dismissed.

Ackerman v. *Protim Services Ltd* shows the severity with which the courts will treat exclusion clauses. They are not generally found in the standard form contracts, but are very commonly used by specialists in an attempt to cut down or limit the duty which would otherwise be imposed by law. Very few of the cases have involved the construction industry, but the principles which emerge from general litigated cases apply with equal force. Some of the cases are inconclusive; *Ackerman* v. *Protim* is not.

An exemption clause will always be interpreted narrowly and against the party putting it forward. Because of the courts' hostility to limitations of liability, they also apply various other interpretative devices to limit the application of a clause. For example, if a clause purports to limit liability for negligence (so far as that is permissible since the Unfair Contract Terms Act 1977), it can only do so if negligence is referred to expressly: *White* v. *John Warwick & Co. Ltd* (1953).

Another group of cases which show the courts' reluctance to give effect to exemption clauses involve clauses which limit the compensation payable for a breach of contract, usually by excluding liability for consequential loss.

This was the situation in *Croudace Construction Ltd* v. *Cawoods Concrete Products Ltd* (1978), in which suppliers were held liable to main contractors for loss of productivity, extra costs of delay in executing the main contract, and a sub-contractor's delay claim, despite a term in the supply contract which said:

'We are not under any circumstances to be liable for any consequential loss or damage caused or arising by reason of late supply or any fault, failure or defect in any materials or goods supplied by us or by reason of the same not being of the quality or specification ordered or by reason of any other matter whatsoever.'

The Court of Appeal held that none of the items claimed could be regarded as 'consequential'; all the losses claimed resulted directly and naturally in the ordinary course of events from the supplier's alleged breach.

Commercially, of course, those who use exemption clauses justify them on economic grounds, i.e. that if the supplier has to accept full liability, the purchaser will have to pay a higher price. This proposition is debatable. What is clear is that despite judicial and legislative hostility, a great deal of construction industry business is done on terms which purport to exclude or limit the supplier's or specialist's liability.

Chapter 3

Architect's Liabilities

Role of the architect

Architects will find a lot of useful information by a careful reading of the judgment in *London Borough of Merton* v. *Stanley Hugh Leach Ltd* (1985), since it gives valuable guidance on their role and functions in administering a building contract.

The arbitrator held that a term should be implied into the contract to the effect that the architect would administer the contract in a proper and efficient manner. The judge agreed with him, and stated that this implied term was:

> 'no more than a restatement or particular application of Merton's contractual undertaking that there would at all times be a person answering to the definition of "the architect" in clause 3 of the Articles of Agreement, and that the architect would be a reasonably competent person who would use that degree of diligence, skill and care in carrying out the duties assigned to him under the contract that could reasonably be expected of an architect appointed to that position'.

This broad statement reinforces the ruling of His Honour Judge John Newey QC, now upheld by the Court of Appeal, to the effect that it is a breach of contract by the employer if he fails to reappoint an architect, should the original architect drop out of the picture: *Croudace Ltd* v. *London Borough of Lambeth* (1986). A term to that effect will be implied if the contract contains no express provision for re-appointment. Employers must therefore take care in choosing an architect, and also ensure that he carries out his duties efficiently and properly.

Merton's central contention in the litigation was that the architect was not, under the contract, the servant or agent of the building owner. Rather, they said, he is introduced into the contract 'solely to hold the ring between the building owner and the contractor'.

This rather surprising contention was firmly rejected by Mr Justice Vinelott. Under JCT terms, he observed:

'the architect acts as the servant or agent of the building owner in supplying the contractor with the necessary drawings, instructions, levels and the like and in supervising the progress of the work and in ensuring that it is properly carried out'.

So far as the architect's discretionary powers are concerned, these are to be exercised with due regard to the interests of both contractor and employer. 'The building owner does not undertake that the architect will exercise his discretionary powers reasonably,' said the judge, but 'he undertakes that although the architect may be engaged or employed by him, he will leave him free to exercise his discretions fairly and without improper interference by him.' And the architect is under a duty to act fairly as between the parties.

Thus far we have effectively nothing more than a welcome restatement of existing law, but the *Leach* case carries us further, because the arbitrator went on to hold that the application of the implied term and the efficient and proper administration of the contract required three things:

- Every drawing issued by the architect should be accurately recorded by means of an AI sheet or some similar procedure.
- Every drawing issued by the architect should have a unique reference.
- Amended drawings should be clearly identified as such and the fact of the amendment should be indicated by the reference.

On this last point, the arbitrator had held that it would be over-precise to require too much detail, but emphasised that the fact of amendment should be indicated by an addition to any reference number or cipher.

Mr Justice Vinelott – while doubting whether this finding was appropriate for review as a preliminary issue – saw no grounds for expressing any different view from the arbitrator. The emphasis for architects is, therefore, on a tightening up of contract administration.

Architects may argue that this ruling places too heavy a burden on the architect, but it seems self-evident that normal practice and procedure requires that the issue of each drawing should be accurately recorded. As the arbitrator said in his award, quoted by the judge:

'such a requirement is necessary for business efficacy as, if there were no such record the only result could be of uncertainty in the minds of the architect and contractor leading to the possibility of endless disputes. The basic contention is therefore an example of efficient administration by the architect.'

Like the arbitrator, I would have thought that this was quite normal, and the fact that architect's drawings must bear a unique reference number is nothing out of the ordinary.

There are, of course, many and varied practices and procedures followed by architects, because no two professional firms conduct themselves in an identical manner. Indeed, as the judge remarked:

'it is clearly impossible to define accurately and comprehensively the various acts of an architect which together make up efficient and proper administration.'

But *Leach* emphasises that inefficient administration by the architect can amount to a breach of contract. Poor contract administration by the architect can easily result in substantial losses to the contractor, and fairness dictates that the contractor must have a remedy. The vast majority of architects is administratively competent, and if those in doubt follow the guidance given in such books as the Aqua Group's *Contract Administration*, 7th edition, (BSP) there should be no problem.

London Borough of Merton v. *Stanley Hugh Leach Ltd* is a very significant case in more ways than one, but this part of the judgment should be studied in depth by architects and consultants. Other aspects of the case (discussed elsewhere) are of more interest to contractors. Certainly, the very clear ruling about the architect's role and status as agent of the employer needs to be emphasised, and contrasted with his other role as independent referee.

No duty of care on architects

In the exercise of his duties under a building contract, the architect owes to the employer a duty to carry out his functions in a professional manner and with due care and skill. This sensible proposition follows from *Sutcliffe* v. *Thackrah* (1974) where a building owner sued his architect for damages for negligence and breach of duty in supervising building work, and for certifying work not properly done by a contractor who subsequently became insolvent. The employer was held entitled to recover against the negligent architect, who could not shelter behind the plea that he was a 'quasi-arbitrator'.

Following that decision of the House of Lords, it seemed likely that eventually it would be held that the architect owed a duty to the contractor in tort to exercise reasonable skill and care when issuing certificates and granting extensions of time. In *Lubenham Fidelities & Investment Co. Ltd* v. *South Pembrokeshire District Council* (1986), His Honour Judge John Newey QC concluded that the negligent architects owed a duty of care to the contractors as well as to the employer. His view received some apparent support from Lord Justice May in the Court of Appeal, although the court did not consider whether or not a duty relationship in fact existed since the issue which concerned the negligence of the architect was not in dispute. In *Shui On Constructions Ltd* v. *Shui Kay Co. Ltd* (1985), the Hong Kong Supreme Court ruled that the architect *was* under a duty to the contractor to act fairly and professionally when acting under the contract and might be liable if he did not.

The point was directly at issue in *Michael Sallis & Co. Ltd* v. *Calil & William F. Newman Associates* (1988), a masterly judgment of His Honour Judge James Fox-Andrews QC, holding that the architect under a JCT contract owed a duty of care to the contractor when certifying and dealing with extensions of time and the ascertainment of direct loss and/or expense. He considered all the relevant authorities and concluded that the contractor had a right to recover damages against an unfair architect.

He said:

'It is self-evident that a contractor who is a party to a JCT contract looks to the architect to act fairly as between him and the employer in matters such as certificates and extensions of time. Without a confident belief that the reliance will be justified, in an industry where cash flow is so important to the contractor, contracting would be a hazardous occupation. If the architect unfairly promotes the employer's interest by low certification or merely fails properly to exercise reasonable care and skill in his certification, it is reasonable that the contractor should not only have the right against the owner to have the certificate reviewed in arbitration, but should also have the right to recover damages against the unfair architect ... To the extent that the plaintiffs are able to establish damage resulting from the architect's unfairness in respect of matters in which under the contract the architect was required to act impartially, damages are recoverable and not too remote'.

Unfortunately, this sensible approach was rejected by the Court of Appeal in *Pacific Associates Inc.* v. *Baxter* (1988), on the rather doubtful basis that it overlooked the contractual structure against which any reliance placed by the contractor must be based and that it was not 'just and reasonable' to superimpose duties beyond those freely agreed in the contractual matrix. Their lordships have told us that there is no simple unqualified answer to the question of whether an architect owes a duty of care to the contractor in tort. 'The question can only be answered in the context of the factual matrix, especially the contractual structure against which such a duty is said to arise': Lord Justice Purchas.

The *Pacific* case is really only explicable because there has been a dramatic policy reversal in the area of liability for purely economic loss and a re-drawing of the boundaries between contract and tort. The case arose under the FIDIC international civil engineering conditions of contract and the contract contained a special clause (PC 86) which provided that 'neither the engineer nor any of his staff shall be in any way personally liable for the acts or obligations under the contract'. This disclaimer was held to be effective, even though it was contained in a contract to which the engineer was not a party. Lord Justice Purchas said:

'In accepting the invitation to tender with the complete contractual framework including the disclaimer in PC86, it is impossible to support the contention that either the engineer was holding himself out to accept a duty of care with the consequential liability for pecuniary loss outside the provisions afforded to the contractor under the contract, or that the contractor relied in any way upon such an assumption of responsibility on the part of the engineer in any way to extend his rights'.

If that were the only basis for the decision it would not be so disquieting. However, Lord Justice Purchas was also of the view that:

'even if PC86 were not included in the contract in this case, the provisions of GC67 (the arbitration clause) would be effective to exclude the creation of any direct duty upon the engineer towards the contractor'.

This was also the opinion of the second member of the Court of Appeal, Lord Justice Russell, who also emphasised the sufficiency of the contractor's contractual remedies. His lordship thought that 'the very existence of clause 67 as drafted was sufficient to dispose of this appeal'. The presence or absence of any arbitration clause providing the contractor with remedies for non- or under-certification may therefore make all the difference to the liability of the certifier.

His lordship went on to make what those with practical experience of building and civil engineering contracts may well consider to be a somewhat surprising comment:

'The contractors in reality had their rights adequately protected by the terms of their bilateral contract with the employer. If they had thought not, then they were at liberty to insist upon a tripartite contract before embarking upon the work'.

Possibly this is the position in theory. However, in the real world of contracting, to imagine that the contractor can negotiate a contract to which the architect (or engineer) is a party is totally divorced from reality.

Pacific Associates Inc. v. *Baxter* is plainly a policy decision in wake of the judicial back-tracking in the field of negligence liability. The conclusion that the engineer owed no duty of care to the contractor depended on the particular circumstances of the case, as Lord Justice Purchas emphasised, and in particular the contractual structure. In *Arenson* v. *Arenson* (1977), Lord Salmon dismissed a submission that there should not be a duty owed by the architect to the contractor because it would put the architect in risk of being 'shot at from both sides', although he did not specifically consider the question of a duty arising between architect and contractor.

In the *Pacific* case, the contractors relied on Lord Salmon's dismissal of the 'shot at from both sides argument'. They argued unsuccessfully that if there was a duty in tort owed by the engineer to the employer, so there ought also to be a similar duty owed by the engineer to the contractor. In the Court of Appeal's opinion:

'this submission is flawed because it ignores the contractual relationship between the employer and the engineer as opposed to the relationship which can only be constructed in tort arising out of the contractual structure accepted by the contractor to determine the duty, if any, owed by the engineer to the contractor'.

In cases of pure economic loss, the law may impose a duty to take care although the responsibility has not been assumed voluntarily provided three requirements are satisfied: foresight of harm, proximity, and that:

'it is just and reasonable to impose the duty, taking into account the nature, extent and likelihood of any harm, foreseeably resulting from any want of care'.

It was on this third point that the contractors failed. Even had there been no disclaimer, Lord Justice Purchas said that:

'it would not be just and reasonable to impose liability. Such a duty would cut across and be inconsistent with the structure of relationships created by the contracts into which the parties had entered, including the machinery for settling disputes'.

Lord Justice Russell took the same line, adding that 'it is not just and reasonable that there should be imposed upon the engineers *a duty which the contractors chose not to make contractual*'. [Author's emphasis.]

The decision puts the legal clock back by 20 years, and clearly the last word has not been said. Lord Justice Purchas noted that:

'the position might well have been otherwise if GC67 or some other provision for arbitration had not been included in the contract. The position of the engineer, might well be arbitral or quasi-arbitral ... On the other hand, he may well owe a duty to exercise skill and care to a contractor who relies on his valuations and certificates ... or even to be shot at by both sides as envisaged in *Arenson* v. *Arenson*. These are all exceptional cases. It will be rare for a contract for engineering works of any substance in which a consulting engineer is appointed not to have an arbitration clause'.

There, for the time being, matters must rest, and architects and engineers have won this round. The more realistic approach of Judge Fox-Andrews in *Sallis* v. *Calil* is no longer to be the law. Perhaps in the not too distant future the House of Lords will restore the law to what many of us think it ought to be; at present there is little likelihood of that happening.

Design is the architect's domain

Although there are a few cases which support the proposition, the bizarre contention that a main contractor owes a duty of care to warn a negligent architect of his own bad design was resoundingly rejected by His Honour Judge Peter Bowsher QC in *University of Glasgow* v. *Whitfield* (1988), where the main contractor was joined as third party by the defendant architect. The JCT 63 contract, with Scottish

Supplement, incorporated detailed Bills of Quantities describing the works and it was clear that the contractor had no design liability under the contract.

Glasgow University complained that the architect was negligent in his design of the Hunterian Art Gallery. The works were practically completed in June 1978, and from August 1978 the Gallery was plagued with leaks and condensation. The main complaint was that the architect had failed to design remedial works with care and skill and that he had given negligent advice when water was seen in the main gallery in 1982 and 1983. In fact, no vapour barrier had been specified and the contractor followed the architect's design.

The judge held that the architect was under a duty of care to the University both in contract and in tort, requiring the reasonable skill and care of the ordinarily competent architect. His designs ought to have taken into account Glasgow's climatic conditions and have had due regard to 'the particular needs and requirements of a building intended for the housing and display of valuable works of art.' The building is on an exposed hillside. His Honour found that the defendant was also negligent in some of the remedial designs and variations from them.

Even if the University's claim was statute-barred in contract it was not in tort because the architect was under a continuing duty to design and review the design. The condensation and consequent waterstaining of structural timber was not 'damage' so as to bar the claim in tort. The judge held that there was no actual damage to the structure of the main gallery until March 1981. The writ was issued in time, and the claim was not barred.

The point about the continuing duty to design is an important one. Judge Bowsher, following the Court of Appeal in *Brickfield Properties Ltd* v. *Newton* (1971), said that 'the duty to design is a continuing duty which extends until the building is complete.' Having considered the various cases he rejected the view that the plaintiff's cause of action accrued on handover, i.e. practical completion. He quoted with approval Sir William Stabb's statement that 'the subsequent discovery of a defect in the design ... reactivated ... the architect's duty': *Merton LBC* v. *Lowe* (1981).

'Where, as here, an architect has drawn to his attention that damage has resulted from a design which he knew or ought to have known was bad from the start, he has a particular duty to his client to disclose what he had been under a continuing duty to reveal, namely what he knows of the design defects as possible causes of the problem.'

Judge John Newey's decisions in *EDAC Ltd* v. *Moss* (1984) and *Manchester University* v. *Wilson* (1984) were not authorities for the argument that the architect's liability ends at practical completion. The judge said:

> 'I can see no reason, in principle, why the duty should be so limited in time despite the fact that the architect's right to require work to be done alters at that point'.

In *Moss*, the court was only concerned with the question whether the duty extended as far as practical completion. In *Wilson*, Judge Newey left open the question whether the duty extended past practical completion. In Judge Bowsher's view it does.

Alternatively, he held that the architect undertook a fresh duty to reconsider the design when he gave advice and charged a fee for giving it in 1982. The plaintiffs were entitled to judgment for the architect's negligence.

After considering other matters, the judge turned to the vital question of the contractor's duty to the defendant architect who had brought him in as third party, seeking contribution from him on the basis that the contractor had a duty to warn of design defects. In sum, since no danger to persons or other property was alleged, the architect was unable to rely on any liability in tort owed by the contractor to the employer, whether for alleged bad workmanship or an alleged duty to warn the employer of bad design. That covered the claim for contribution. But did the contractor owe a duty to the architect himself?

The damage, which the architect alleged the contractor should have guarded him against, was his being found liable to pay damages to the employer as a result of his own defective design! That was pure financial loss and was irrecoverable because no duty was owed to the employer in respect of it. There was no 'special relationship' and the point was covered by *Simaan General Contracting Co.* v. *Pilkington Glass Ltd* (1988).

The decisions of Judge Newey in *Moss* and *Wilson* were of no assistance to the architect. Both of them were concerned with a duty of a contractor to warn the employer and not with a duty owed to the architect to warn the architect himself. The judge's references to a duty to warn the architect were to the architect *as agent of the employer* and, in *Wilson*, Judge Newey made it clear that both his decisions were founded on an implied contract between the contractor and the employer. In both cases he had cited *Duncan* v. *Blundell* (1820) and *Brunswick Construction Co.* v. *Nowlan* (1974), where the architect provided the design but took no further part in the project, so the contractor knew that the owner placed reliance on him for the design.

In Judge Bowsher's view *Moss* and *Wilson* were to be read as cases where there was a 'special relationship' between the parties. Here, the contractor was not being relied on for design advice and had no reason to suppose that he was. Design, and responsibility for it, was the architect's province and not the contractor's.

The judgment closed with a warning, in that there are some circumstances in which a term might be implied or a duty owed in tort which required a contractor to warn a building owner of defects in the design, and contractors should read Hudson's *Building Contracts*, 10th edition, p. 291, for a review of the possibilities.

The powers of the architect

The architect's powers under JCT 80 are limited. He can only do those things which the contract authorises him to do. Thus, clause 4.1, which obliges the contractor to comply with architect's instructions, refers specifically to instructions 'in regard to any matter in respect of which the architect is expressly empowered by the Conditions to issue instructions'.

In contrast, under ACA 1982 terms, clause 1.1 requires the contractor to 'comply with and adhere strictly to the architect's instructions *on any matter* whether mentioned in this Agreement or not'. This must, however, be read subject to other provisions of the contract, such as clause 9.3, which says 'where the contract documents provide ...', and which also have a limiting effect.

The power of the architect is also limited at common law; he cannot instruct the contractor *how* to do his work: *Clayton* v. *Woodman and Son Ltd* (1962).

In some cases architects do act in excess of their authority and a reading of the reports of the long-running saga of *Stockport Metropolitan Borough Council* v. *William O'Reilly* is instructive. The second instalment is reported in [1983] 2 Lloyd's Rep 70, where Mr Justice Neill gave leave to Stockport to revoke the authority of the arbitrator and made various consequential orders. He declined to remit the whole case to the arbitrator to look at again.

The earlier proceedings dealt with the architect's powers. O'Reilly contracted to build 105 houses, garages and ancillary works for the council, the contract being in JCT 63 form. The job went sour and various disputes between the parties were referred to arbitration.

Essentially, these concerned (a) the extent to which the works were varied and whether the variations were authorised, and (b) the termination of the contract, the council contending that it had been properly determined and O'Reilly arguing that it had repudiated the contract.

The arbitrator's interim award of 11 November 1976 was long. He seemed to confuse variation of works with breach of contract, failing to distinguish between those acts of the architect for which the employer was responsible and instructions and orders made in excess of his authority.

The matters which led the arbitrator to his conclusions (which were substantially in favour of O'Reilly) could be put in three groups: (a) faults of the employer; (b) faults of the architect as agent for the employer – which were legitimate factors in deciding who repudiated the contract; and (c) faults of the architect otherwise than as agent of the employer, which were not.

In the event, His Honour Judge Edgar Fay, QC, set the arbitrator's interim award aside, and provided an admirable summary of the law relating to an architect's authority. He said:

> 'An architect's *ultra vires* acts do not saddle the employer with liability. The architect is not the employer's agent in that respect. He has no authority to vary the contract. Confronted with such acts, the parties may either acquiesce, in which case the contract may be *pro tanto* varied and the acts cannot be complained of, or a party may protest and ignore them. But he cannot saddle the employer with responsibility for them.'

There is an important practical point under JCT 80, because clause 4.2 enables the contractor to challenge the architect's authority if he receives an instruction which he suspects is outside the architect's powers. In short, the contractor can ask the architect to specify his contractual authority for issuing the alleged instruction by following the procedure laid down in the sub-clause.

> 'The architect shall forthwith comply with any such request and if the contractor shall thereafter comply with the said instruction ... the issue of the same shall be deemed for all the purposes of the contract to have been empowered by the provision ... specified by the architect in answer to the contractor's request.'

There is no analogous provision in the ACA Contract but, as I have indicated, there are legal limitations on what the architect is authorised to do. The employer cannot deny that the architect has his actual authority to do those things specified in the contract, and it is not for the contractor to consider whether or not the architect does have the actual consent of the employer for instructions issued within the terms of the contract.

The arbitration proceedings in *Stockport Metropolitan Council* v. *O'Reilly* were begun as long ago as February 1971, pleadings being exchanged in the same year, and the case continued on and off for years. For present purposes, its main significance is the matter of the architect's authority, but the case reports should be read by anyone involved in the arbitral process.

Reverting to Judge Fay's judgment, and the point that the employer is not liable to the contractor for acts of the architect not within the scope of his authority, does the contractor have a remedy? Since the architect is not a party to the contract, there is no contractual remedy and only in the rarest cases will the contractor have a remedy in tort against him because of the ruling of the Court of Appeal in *Pacific Associates Inc.* v. *Baxter* (1988) which, unfortunately has clouded the issue.

The contractor may, however, sue the employer as principal and the architect as agent jointly where the actual authority is obscure and, might in some cases recover damages for breach of warranty of authority of an amount such as will indemnify the contractor and put him in the same position as if the architect's act had been within his authority: see Emden's *Building Contracts*, 8th edition, p. 477.

Clearly, there are perils for both architects and contractors under our present system. The main thing is for both architect and contractor to be sure that the necessary authority exists under the contract conditions.

If a 'variation of contract' is needed, i.e. a change in the contract itself, then the contracting parties (the employer and the contractor) must enter into a supplemental agreement, which is a simple device which can overcome many difficulties.

How not to run a contract

The case of *Pincott* v. *Fur & Textile Care Ltd* (1986) should be compulsory reading for all architects since it provides an object lesson in how not to run a contract. Contractors, too, will welcome the decision since the judge expressly ruled that an architect must

'advise his client, and not mislead the contractor, as to the nature of any agreement that he recommends should be entered into between the client and the contractor.'

An architect was commissioned to prepare a feasibility study and later to design and supervise extensive alterations to a north London dry cleaners. Although the architect applied for planning permission, he let the contract in JCT Minor Works form before planning permission was granted. Since much specialist work was involved – electrical, ventilation and air conditioning, for example, it is doubtful whether the Minor Works form was really suitable. The contract sum was £105 563.

Work started in January 1982 but planning permission was refused in July 1982. The architect also failed to obtain other statutory consents or to agree a party wall award with the adjoining owners. The late Judge David Smout QC was unhesitatingly of the view that the architect was guilty of professional negligence and not merely of errors of judgment. He said:

'The duty of care of a competent architect, embraces the need to avoid, so far as reasonably possible, any unnecessary expenditure for the client.'

This presupposes that the architect should:

- Give adequate warning to his client that work undertaken without planning permission or other necessary consents is likely to involve additional expense and possible demolition;
- advise the client, and not mislead the contractor, about the nature of the contract he is recommending be used;
- use his best efforts speedily to obtain all necessary consents;
- notify the client if the required consents are not forthcoming so that other procedures can be adopted.

The architect in question failed to do any of these things:

'He never notified [the employer] that planning permission had been refused, nor indeed of the failure to obtain other consents, and only hinted at such matters in his resignation letter of November 8 1982,'

said the judge.

The judge's finding that the architect was negligent in failing to advise the client about the true nature of the Minor Works contract and in misleading the contractor about his role under the form is one of the most interesting parts of the decision.

The architect had argued that he never intended that the contractor should be responsible for any work other than his own, despite the fact that he had included a sum for specialists' work in an early interim certificate. The learned judge put the matter succinctly:

'In signing (the contract the contractor) thereby undertook to carry out all the works described in the agreement and was in the position of a main contractor, and those who were named in the schedule of costs attached ... were on the face of it intended to be sub-contractors. The truth is that (the architect) was misrepresenting to (the contractor) the nature of the agreement so as to ensure that it was signed.'

Interestingly, the court also held that various certificates issued by the architect under the contract were invalid. By 13 August 1982, so the judge found, the contract machinery had broken down, but on that day, as a tactical device to save something for the specialists involved, the architect issued a certificate of practical completion, and an interim certificate so as to obtain payment for sub-contractors. He made no valuation, but merely put in round figures. The judge remarked:

'The motive may have been commendable, but as the certificate was not intended to represent any genuine valuation I hold that it was not *bona fide* in the sense that it was not issued for the purpose of according with the terms of the contract. It was not valid.'

The judge reached the same conclusion about the architect's certificate granting a retrospective extension of time to the contractor. This he did in response to the contractor's request, which was made after the anticipated date for completion had passed and not before it, which is what the contract requires. Although the judge doubted whether time is of the essence under MW80, clause 6(ii), he declared the architect's grant of a 16 week extension invalid because the architect had responded automatically and had not exercised any independent judgment, as the contract also requires.

Damages totalling nearly £50 000 were awarded, plus professional fees subsequently incurred by the employer in obtaining planning and other consents, although the architect involved recovered his professional fees of some £14 500. Although in money terms *Pincott* v. *Fur & Textile Care Ltd* was a small case, in practical terms it is an important one, and illustrates gaps in an architect's training.

Architectural education in the UK is gravely defective in that there is little legal input and very little emphasis on contract administration. The newly fledged architect is thrown in at the deep end without adequate preparation on the practical side. As this case shows, an architect is expected to have a detailed knowledge of building contracts and contract procedures and cannot soldier on in happy ignorance. Equally, he must have a knowledge of case law developments affecting his professional activities. He is not a mere designer.

So far as standard form contracts go, the position is well-stated in Hudson's *Building Contracts* (10th edition, p. 123):

> 'The time is rapidly approaching when architects ... recommending the use without modification of some of the standard form contracts in general use at present in the United Kingdom ... must be in serious danger of an action for professional negligence.'

That barbed shaft was directed particularly at JCT 63, but the philosophy underlying it remains true.

Impartiality and implied terms

The Australian decision of *Perini Corporation* v. *Commonwealth of Australia* (1964) deals with the important question of the duties of architects and others who act as certifiers under the terms of a construction contract.

In this case the contractor complained that his applications for extensions of time had not been properly considered by the Director of Works who, under a special form of contract, was performing duties equivalent to those of an architect under the JCT form of contract or of an engineer under the ICE form. The contractor's argument was that the certifier was under a duty to act impartially and that it was an implied term of the contract that the employer would direct him to do so.

The statements of principle by the Supreme Court of New South Wales, though not strictly binding on English courts, are most instructive and it is thought that they apply equally to the normal English construction contracts. Under the express contract terms, the certifier was required to make his decision on applications for extension of time 'within a reasonable time'.

Mr Justice Macfarlan had this to say:

> 'The measurement of a reasonable time in any particular case is always a matter of fact. Plainly the [architect] must not delay, nor may he procrastinate, and in my opinion he is not simply entitled to defer a decision. On the other hand he is ... necessarily obliged to have available for that consideration such time as is necessary to enable him to investigate the facts which are relevant to making it.

When that investigation is complete ... the decision should then be made.'

The judge then considered the nature and extent of the terms which were to be implied by law into the contract. His ruling on this matter represents a succinct and accurate statement of the common law position.

Contractual terms are to be implied only when their subject matter is not covered by the *express* terms of the contract. Quoting a pre-war Australian case the judge said:

'Terms are to be implied only when the matter to which they relate is not covered by the express terms of the contract, and if not annexed by usage, statute or otherwise, is such that it is clearly necessary to imply the term, in order to make the contract operative according to the intention of the parties as indicated by the express terms.'

The contractor contended that two terms should be implied into the contract. First, there was a *negative* implied term that the employer should not interfere with the proper performance by the Director of Works of the duties imposed on him by the contract. Second, that there was a *positive* implied term that the employer was bound to ensure that the director properly performed the various duties imposed on him contractually.

Mr Justice Macfarlan found in favour of the contractor on both points. As to the first point he said that, having held that the director acting as certifier was duty-bound to act independently and in the exercise of his own volition:

'it is not possible to assume that the parties to this agreement could have contemplated that he would act in a manner other than that upon which they have agreed and expressed in (the contract) and that it is a consequence of this assumption that they shall have implicitly bound themselves one to the other that they would not do anything that would prevent him from the proper discharge of the mandate which contractually they granted to him ...'

So far as the positive implied term was concerned, the learned judge ruled that the contracting parties are:

'bound to do all cooperative acts necessary to bring about the contractual result. In the case of the (employer) this is an obligation to require the director to act in accordance with his mandate if the (employer) is aware that he is proposing to act beyond it ...'

The significance of this statement is apparent when one considers the question of employed architects – and to all intents and purposes the director was acting in that capacity.

Mr Justice Macfarlan was attracted to the views expressed by Lord Justice Scott in *Panamena Co. Ltd* v *F. Leyland and Co. Ltd* (1943), which was a shipbuilding case, later to go on appeal to the House of Lords. The employer in *Perini* argued that the House of Lords' judgments went against the contractor's contention. However, the judge preferred the views of Lord Justice Scott who said that, in cases of this sort, employers are

'under a contractual duty to keep their (certifier) straight on the scope of what I metaphorically called his "jurisdiction" by which I do not mean that he was in any sense an arbitrator, but only that as an expert entrusted with the duty of impartiality within a certain sphere he had to form his opinion with judicial independence within that sphere ...'

Perini v. *Commonwealth of Australia* is an important decision and the judgment repays careful study. It does not suggest that an employer gives a warranty or undertaking that his architect or engineer will always make the right decision. But in building and engineering contracts there is to be implied a term that the employer will do nothing to prevent the architect or engineer from discharging his duties as certifier. Equally, there is to be implied a term that the employer will do all that he reasonably can to ensure that the architect acts properly in accordance with his powers and exercises them under the contract.

Chapter 4

Performance, Completion and Payment

The effect of a programme

An important judgment was handed down by the High Court in *Glenlion Construction Ltd* v. *The Guinness Trust* (1987), and stills some of the controversy about contractor's programmes which show intended completion before the contract completion date.

The decision deals with questions of general importance to the industry and, in particular, whether in a JCT contract where the contractor is required to provide a programme, there is an implied term that the employer should so perform the contract as to enable the contractor to carry out the work in accordance with the programme and complete the works on the programmed completion date which may be earlier than the contractual completion date.

Although in Glenlion the contract was in JCT 63 form, JCT 80 incorporates many of the relevant provisions and so the ruling is of general application.

The Glenlion contract related to a residential development at Bromley, Kent, and the date for completion was 114 weeks from possession. The Bills contained a 'Progress Chart' requirement: 'Provide within one week from the date of possession a programme chart of the whole works ... showing a completion date no later than ...' The contractor's programme as submitted aimed for completion 101 weeks from the date of possession: he had programmed to complete early. Contractually, the contractor is bound to complete 'on or before' the contract completion date, but it is common practice for contractors to plan early completion.

The first question which His Honour Judge James Fox-Andrews QC had to decide was whether the Bill requirement calling for a programme was a term of the contract. The Bills were a contract document (JCT 63, clause1(1), JCT 80, clause 2.1), but clause 12(1) (JCT 80, clause 2.2.1.) said that 'nothing contained in the Contract Bills shall override, modify or affect' anything contained in the printed conditions – a provision which, in my view, should be struck out of the

JCT conditions, since its literal application leads to many absurdities. In this case it did not matter; the programme requirement supplemented the printed conditions and did not override or modify them. The learned judge ruled that the full wording of the Progress Chart clause was a contract provision. This led to the heart of the case, and two basic questions had to be answered:

- So far as the contractor's programme showed a completion date before the Date for Completion, was the contractor entitled to carry out the works in accordance with his programme and so complete early?
- Was there an implied term in the contract to the effect that the employer and his architect should carry out their obligations so as to enable the contractor to complete early as he had planned?

The contractors conceded that builders frequently produce programmes that are over-optimistic, and the judge inferred that possession of the site was given to Glenlion before the contract was signed. This was not relevant to interpreting the contract even though, on this assumption, the employers knew before signing the contract that early completion was planned.

In light of the wording of JCT 63, clause 21, (JCT 80, clause 2.3.1.) obliging the contractor to complete 'on or before' the contract completion date, Judge Fox-Andrews held that:

'it is self-evident that Glenlion was entitled to complete before the Date for Completion.

And the contractor was entitled to complete on an earlier date whether or not he produced a programme with an earlier date and whether or not he was contractually bound to produce a programme'.

This commonsense interpretation knocks a common architect's argument on the head and, what is more, the judge added:

'If [the contractor] was entitled to complete before the date for completion, he was entitled to carry out the works in such a way as to enable him to achieve the earlier completion date whether or not the works were programmed.'

But Glenlion fell at the final hurdle. To succeed, it had to satisfy the stringent tests for the implication of a contractual term. Glenlion did not suggest that it was both entitled *and* obliged to finish by the earlier completion date. If there was such an implied term it imposed an obligation on the employer but none on the contractor.

As the judge remarked, it is unclear how the variation provisions would have applied in those circumstances. The extension of time provision only operates in relation to the Appendix Completion Date. He said:

'A fair and reasonable extension of time for completion of the works beyond the Date for Completion stated in the Appendix might be an unfair and unreasonable extension from an earlier date'.

Judge Fox-Andrews thought that it was not immediately apparent why it would be reasonable or equitable to place an absolute unilateral obligation on the employer. He referred to textbook authorities, and in particular to the Supplement to Keating's *Building Contracts* 4th edition dealing with optimistic programmes submitted to the architect.

'It is then argued that the contractor has a claim for damages for failure by the architect to issue instructions at times necessary to comply with the programme. While every case must depend upon the particular express terms and circumstances, it is thought that, upon the facts set out, the contractor's argument is bad ...'

The learned judge agreed. 'The fact that the programme is required to be provided by the contractor does not in itself make the position different,' and so Glenlion lost the day on this important point.

'It is a foregone conclusion that Glenlion cannot establish that such a term was so obvious that it went without saying. The contract as drawn is efficacious in the sense that it produces ... the desired effect.
  The unilateral imposition of a different completion date would result in the whole balance of the contract being lost. The position would be no different if the obligation imposed on the [employer] was, instead of being absolute, a requirement that the [employer] should act reasonably.'

Contractor's dual obligation

Some standard form construction contracts require the contractor to execute the work according to a specification *and* to the satisfaction of the employer's engineer or architect. It is sometimes argued that, in general, building contracts are to be interpreted differently from other contracts, so that this phrase places a single obligation on the contractor, rather than a cumulative one.

That view was rejected decisively by the High Court in *National Coal Board* v. *Wm Neill & Son (St Helens) Ltd* (1984) where the point arose under the BEAMA RC Conditions (1956 edition). Clause 4(1) of that form provided that:

> 'All plant to be supplied and all work to be done under the contract shall be manufactured and executed in the manner set out in the specification, if any, and to the reasonable satisfaction of the [NCB's] engineer.'

The defendant carried out works at a Lancashire colliery. The engineer issued a certificate of satisfaction with the work. The NCB paid the contractor. Two years later one of the gantries carrying two of the conveyors collapsed. The NCB alleged that this was due to the contractor's failure to execute the work in accordance with the specification and claimed damages for breach of contract.

The contractor denied liability. It argued that there was a general rule in building contracts that clauses in terms similar to clause 4(1) should be interpreted merely as imposing an obligation to execute the work to the satisfaction of the employer's engineer. Such clauses should not be interpreted as imposing on contractors a cumulative obligation to execute the work both in the manner prescribed by the employer *and* to the satisfaction of the engineer.

The contractor argued that once the engineer had certified his satisfaction with the work, his certificate was *conclusive* evidence that the work had been carried out satisfactorily and that once this had been done the employer could not complain of any failure to comply with the specification.

That argument was rejected by the High Court. There is no general rule of the sort the contractor maintained.

The general principle of interpretation which applies to *all* contracts is a simple one. In each case the meaning of any clause in a contract has to be ascertained by looking at the contract as a whole and giving effect – so far as possible – to every part of it.

The judge ruled that, on the true interpretation of clause 4(1) of the BEAMA RC Conditions, the contractor was under a dual obligation. First, it had to execute the work in the manner set out in the specification. Secondly, it must execute the work to the reasonable satisfaction of the engineer. It is, therefore, a dual obligation – and this principle is clearly of the greatest importance.

The previous authorities are unsatisfactory, and the court's judgment attempts to reconcile them. The judge took the view – very sensibly – that where the engineer (or architect) issues an unqualified certificate expressing his satisfaction, that certificate was not an irrebuttable presumption of law, even though 'in practice the architect or engineer would be unable to rebut it'. But this is a question of fact in each and every case.

Under JCT contracts the arbitration clause provides that an arbitrator shall have power to open up and review certificates so that the certificates are not intended to be conclusive, except to the extent that the contract provides, i.e., the conclusive evidential effect given to the final certificate. In the BEAMA RC form the arbitration clause did not enable the arbitrator to open up certificates.

The general principle of interpretation is that the contract must be read as a whole – a point often forgotten when attempting to make one's point in argument. What really appeared decisive, in the judge's mind was the judgment of the High Court in *Billyack* v. *Leyland Construction Co. Ltd* (1968), where it was held that a cumulative obligation should always be adopted in respect of two obligations joined by the word 'and'. The problem there was that in that case the satisfaction was to be that of an independent third party (a building inspector) and not the agent of the owner.

Clause 3 of the BEAMA RC Contract also provided for the engineer's approval of the contractor's drawings. The judge said:

> 'It would be a strange interpretation of this clause to read it as releasing the [contractor] from express warranties contained in the specifications and tenders. There are many matters which a reasonably competent engineer would be unable to check, at least without great difficulty and expense'.

To check the strength of the structures designed by the contractor, for example, the engineer would have in effect to repeat all the work himself. This was not, in the judge's view, the extent of the engineer's duty.

The older cases are of little help today in this context. A century ago, building was a much more straightforward operation, requiring no great specialist skills other than those ordinarily to be found in the local builder, and an architect then could be expected to have at least the expertise of the ordinary builder in every aspect of the craft.

'The same may still apply in the case of ordinary houses,' said the judge, but not today in general 'where specialised methods of construction are involved and usually several specialist sub-contractors on any one contract.'

In the judge's view, an architect or engineer cannot be expected to be an expert in every aspect of specialist trades and must rely in part on the specialist's expertise. This is one of the most interesting aspects of the judgment and should put contractors on their guard.

'Due diligence and expedition'

The Court of Appeal decision in *Greater London Council* v. *Cleveland Bridge & Engineering Co. Ltd* (1986) cleared up a doubtful point of great importance.

A contractor is not obliged to so conduct his programme and commencing dates as to afford the maximum advantage to the employer. His obligation is to complete the works within the time limit of the contract.

This is so even if there is a clause requiring the contractor to work with 'due diligence and expedition' because such a provision must be read in light of the other contractual provisions about the date of completion.

The trial judge, whose decision was affirmed, said:

> 'As it is a general principle of building contracts that it is for the contractor to plan and perform his work as desired within the contractual period, it cannot be said that [he has here] failed to exercise due diligence and expedition'

because he had complied with the overall deadline.

The litigation arose out of the construction of the Thames Barrier. Cleveland Bridge contracted with the GLC to manufacture and erect the gates and gate arms, and a dispute arose about the application of the price-adjustment clause which was based on the BEAMA formula. The GLC contended that it had overpaid Cleveland some £3.5 million on the basis that there was an express or implied term that the contractor would proceed with due diligence and was in default even though Cleveland actually completed the work on time.

Clause 19 of the contract provided that 'if the contractor shall neglect to execute the works with due diligence and expedition ...' the GLC was entitled to terminate the contract after giving due notice. The fluctuations clause disentitled the contractor from recovering any price increases resulting from his default or negligence.

Cleveland completed the totality of its obligations in time, but the

GLC argued that the contractor was also bound by an additional obligation 'to accomplish manufacturing and other tasks with due diligence and expedition.'

Lord Justice Woolf put the matter neatly in his judgment.

'the GLC seeks to establish that an obligation exists either as a matter of implication or as an implied term in order to protect itself from increased costs by treating the contractor as being in default or as being negligent ... if the contractor so conducts his manufacturing process so as to unnecessarily increase the amount payable to [him] by the manufacture of the works, although [his] ... performance in fact complies with his express obligations under the contract.'

This contention found no favour with either the trial judge or the Court of Appeal. None of the clauses of the contract obliged the contractor so to conduct his programme of work as to afford the best advantage to the GLC.

There was no possible basis on which an implied term could be written into the contract, which was commercially effective without it.

The determination provision – clause 19 – did not impose a positive obligation on the contractor to execute the works with due diligence and expedition. It merely provided a remedy if it failed to do so and it was a remedy of which the GLC had not availed itself. The trial judge put it this way:

'One can readily understand that the employers might not wish to wait until the completion date before they have a remedy against a dilatory contractor and might not be able to assert, with conviction, before then that he had wholly and finally disenabled himself from completing the work in time, so as to be able to rely on the common law remedy for anticipatory breach. But the clause resolves that problem because the employer may discharge the contractor if he fails to execute the works with due diligence.'

Although the contractor's failure to proceed with due diligence and expedition might entitle the employers to terminate the contract, 'it would not by itself be a breach of contract' by him, the judge said.

The Court was also unable to accept the GLC argument that the delay by the contractor amounted to 'default or negligence' so as to disentitle it to benefit from the fluctuations clause. As between the immediate parties to the contract, 'default' has no wider meaning than the term 'breach of contract'.

Lord Justice Woolf gave several interesting reasons for finding against the GLC:

- The fluctuations clause was designed to protect the contractor, and not the employer, against increased costs.
- If the contractor was unduly to expedite completion it was easy to envisage circumstances when it would not work to the GLC's benefit, because it would then have to pay for works earlier than it might wish. That, at least in theory, could involve the council in substantial extra costs by way of storage.
- Once the contractor had failed to fulfil his express contractual obligations with regard to performance, he would certainly be in default and could not then take advantage of the fluctuations clause.
- Other contract terms provided the GLC with adequate remedies to deal with abuses, e.g. clause 19 itself, entitling them to discharge the contractor.

In effect, as another member of the Court of Appeal pointed out, the GLC was saying that if a contractor has four years to complete a project which could take as little as 10 months and he begins the job at the beginning of the contract period, he is, nevertheless, obliged to complete it within ten months. This is not what the contract said and the contractor was not in breach of any contractual duty. Cleveland was under no obligation to programme its work to ensure that the fluctuations clause provided the least benefit to himself and the maximum benefit to the employer. Such a suggestion was unworkable.

GLC v. *Cleveland Bridge & Engineering Co. Ltd* settled a long-running argument as to the extent of the contractor's performance obligations.

Practical completion

'Practical completion' is an ambiguous concept, and the wording of JCT 80, clause 17.1 does nothing to clarify the ambiguity. The ACA Agreement (1982) side-steps the issue, by using the phrase 'fit and ready for taking-over' in clause 11.1.

In general terms, practical completion may be said to take place when the building is reasonably safe and not unreasonably inconvenient, and in practice there is seldom real disagreement between contractors and architects as to whether or not works are 'practically complete.'

Legal commentators, however, are not so certain. Donald Keating

(*Building Contracts*, 4th edition, p. 324), in discussing the phrase 'when ... the Works are practically completed ...' in clause 15 of JCT 63, says that no clear answer can be given as to the meaning of the phrase and, despite the change of wording in clause 17.1 of JCT 80 the ambiguity remains.

The question is whether the phrase covers the situation where the works are substantially finished but there are minor apparent defects, and since the architect's power to order the remedying of defects during the defects liability period is limited to those defects 'which shall appear' during that period, the answer to the question is more than theoretical.

There are two conflicting views, both expressed in the House of Lords, more recently His Honour Judge John Newey QC grappled with the problem in *H. W. Nevill (Sunblest) Ltd* v. *William Press and Son Ltd* (1982).

In that case, the contract was in JCT 63 form, and there were several points at issue, including, *inter alia*, whether the employer's remedies in respect of the alleged defective work were limited to the remedies specified in clause 15 of the contract (now JCT 80, clause 17), to which the answer was, not surprisingly, in the negative.

In the course of the argument, the contractors made two main submissions. First, that they were not obliged to achieve 'perfection' by the completion date, because the contract envisaged that 'defects, shrinkages or other faults' might occur during the defects liability period, and, therefore, defects and the like did not amount to breaches of contract; and also that they had a right to re-enter the site and to make good defects.

Alternatively, if the defects constituted breaches of contract, clause 15 limited their consequences. The contractor was obliged to return to the site to make good and the employer was obliged to allow it to do so. The risk of consequential loss was borne by the employer.

The learned judge rejected both these contentions. He took the view that clause 15 did not extend the contractor's time to finish the works correctly, nor limit the employer's right to possession of its own site. All that the contractor had was a licence to occupy the site up to the date of completion, and clause 15 only gave a right to re-enter to such extent as was necessary to remedy defects pursuant to the defects schedule or architect's instructions.

He further held that defects in the drains discovered after the issue of the certificate of practical completion undoubtedly constituted breaches of contract. The employer was also entitled to claim damages for consequential loss.

In reaching this conclusion the judge considered the meaning of the words 'practically completed'. In *J. Jarvis and Sons Ltd* v. *Westminster Corporation* (1970), Lord Dilhorne took the view that practical completion meant that there must be no defects apparent in the works at the date which the architect issues a Certificate of Practical Completion.

'The Defects Liability Period is provided in order to enable defects not apparent at the date of practical completion to be remedied. If they had been then apparent, no such certificate would have been issued.'

This was in contrast to the views expressed by Lord Diplock in *P. and M. Kaye Ltd* v. *Hosier and Dickinson Ltd* (1972), which suggested that the architect was entitled to withhold his certificate until all known defects, except trifling ones, were corrected.

In other words, if the *Jarvis* view, is correct, 'practically completed' means that there must be no substantial apparent defects, and that the works can be practically completed notwithstanding that there are latent defects and, indeed, *minor* patent defects which can easily be put right without loss to the employer during the defects liability period.

In *Nevill* v. *Press*, the learned judge favoured the *Jarvis* interpretation and his decision appears to be in accordance with both law and justice. He said:

'I think that the word "practically" in clause 15(1) gave the architect a discretion to certify that William Press had fulfilled its obligation under clause 21(1) where very minor *de minimis* works had not been carried out, but if there were any patent defects in what William Press had done the architect could not have given a certificate of practical completion.'

In straightforward terms, under a JCT contract, the architect is quite justified in issuing his certificate if he is reasonably satisfied that the works accord with the contract, notwithstanding that there are very minor defects which can be remedied during the defects liability period. There seems to be no difference under JCT 80 on this point, despite the change in wording, and the contrary view (founded largely on Lord Diplock's observations in *Kaye* v. *Hosier and Dickinson*) seems no longer tenable. The consequences of the issue of the certificate of practical completion are very important, as regards liquidated damages, rentention, insurance, and so on.

'Practical completion' is not necessarily synonymous with 'substantial completion' as discussed in the law books. Architects who are still doubtful may follow Donald Keating's advice and 'obtain a written acknowledgement of the existence of, and an undertaking to put right, the defects from the contractor.' This, as he says, will ordinarily protect the employer's interests adequately, and is consistent with the intention behind clause 17.

The right to remuneration

A *quantum meruit* claim is one for a reasonable sum. The Latin tag means, loosely, 'as much as he has earned', and it must be admitted that many so-called contractors' claims for 'loss and/or expense' are submitted on this basis. In other words, the contractor looks at what the job has cost him, allows for profit, and sets this against what he has received. The difference is then the amount of his 'claim' against the employer! That, of course, is a nonsensical approach and is rightly rejected.

There is no English case which supports it, but something near to it was accepted in Canada in *Penvidic Contracting Co. Ltd* v. *International Nickel Co. of Canada Ltd* (1975). In that case, the contractor was held entitled to claim the price he would have charged for the job had he known of the difficulties likely to be encountered at the time he submitted his tender. In fact, the explanation of this decision is that the contractor had performed work which was substantially different from that for which he tendered and undertook to do. The breadth of the standard variation clauses in common use makes a similar ruling here improbable.

A *quantum meruit* situation is neatly illustrated by *Amantilla Ltd* v. *Telefusion plc* (1987). This, like so many interesting Official Referee cases, came up on a procedural point.

The plaintiff carried out building and shop-fitting works for the defendant between November 1978 and May 1979 for an agreed VAT-exclusive lump sum price of £36 626. In February 1979 the plaintiff agreed to carry out extensive extra works on a day-work basis, but no price was actually agreed between the parties. The plaintiff had worked for the defendant before. It was assumed that the extras would be costed on a similar basis as before.

Between November 1978 and 6 March 1979 the plaintiff received interim payments totalling £53 000, but his request for a further £5000 as an interim payment was not met. Subsequently, there were several meetings and a great deal of correspondence between the parties and, at the defendant's request, the plaintiff submitted a detailed breakdown of the cost of the extra works.

There was a lengthy meeting in Birmingham in January 1983 when the defendant's manager expressed satisfaction with the work done. He also confirmed that the defendant would shortly submit an offer of 'somewhere between' £10 000 and £132 000 to settle the account.

Following that meeting, on 3 February 1983, the defendant wrote to the plaintiff referring to the meeting and 'enclosing a cheque for a further interim payment in connection with the work'. The letter asked for further details of one item in the account. These were duly supplied on 28 March 1983. After a further meeting, the defendant offered £2000 in full and final settlement. Perhaps not surprisingly the offer was rejected and the plaintiff issued a writ on 30 April 1986. This was more than six years after completion of the works and, so it was alleged, the plaintiff's claim was statute-barred. The writ claimed £142 347.74 plus interest and the plaintiff sought summary judgment for that amount (or a sum to be assessed), or alternatively, an interim payment to cover actual disbursements of £40 000.

The question before His Honour Judge Esyr Lewis QC, Official Referee, was whether the defendant's letter and payment of February 3 1983 amounted to an acknowledgement or part-payment of 'any debt or other liquidated pecuniary demand' for the purposes of section 29(5) (a) of the Limitation Act 1980. If it did, then the action was not statute-barred because acknowledgement or part payment is sufficient to revive a cause of action.

In a very careful judgment, the judge ruled that it did. The action was not barred by lapse of time. The important point is that the judge found that a claim on a *quantum meruit* for a 'reasonable sum' was a sufficiently certain contractual description to enable its amount to be ascertained by the court. The letter and payment amounted to an acknowledgement within section 29(5) (a), even though it was not an acknowledgement of the amount claimed or any particular amount. Accordingly, he gave judgment for the plaintiff for a sum to be assessed if not agreed.

The comfort in the case is that there will be liability to pay a *reasonable price* where the contract is silent on the matter. Here the agreement was to carry out the extra work on a 'day-works' basis – and this is not a term of art. What was in fact agreed was that the defendant would pay a 'reasonable price'.

As all the textbooks point out, in determining what is a reasonable price the courts will consider evidence of the actual cost, plus a reasonable percentage for profit, or they may receive evidence of what reasonable rates for the work involved are. The position is neatly summarised by Hudson's *Building Contracts*, 10th edition p. 497:

'The decision may depend upon the nature and amount of the work involved, or upon whether the work has actually been carried out, or is hypothetical work, e.g. in a claim for damages for loss of profit, or upon the nature of the issues between the parties, and will be governed by considerations of convenience from the evidentiary point of view. Some work is, of course, by nature incapable of accurate measurement by means of rates.'

Amantilla Ltd v. *Telefusion plc* is merely one illustration of how a *quantum meruit* claim may arise. It will also be appropriate following a breach of contract where the innocent party has been prevented from earning a lump sum (*Planché* v. *Colburn* (1831)), as well as where work is done at the request of the other party but without there being an express contract, as in the 'letters of intent' cases. But the right to reasonable remuneration can only arise in such a case where, from the circumstances, an intention that the work should be paid for can be implied.

But it is *not* appropriate where the contract itself contains machinery for ascertaining the price (as in the case of varied work under the standard forms) nor can it be used to justify an otherwise unsupportable claim for 'loss and/or expense'. Contractors at all levels are best advised to agree to the price before undertaking extra work unless the necessary contractual machinery is there.

The certificate issue

In recent years an increasing number of cases concerning architects' certificates have reached the courts, most of them as a result of the Court of Appeal decision in *Northern Regional Health Authority* v. *Derek Crouch Construction Co. Ltd* (1984). That case decided that under a JCT contract the court has no power to go behind an architect's certificate, which only an arbitrator has power to open up or review: JCT 63, clause 35. Despite criticisms *Crouch* has been followed time and again where the wording of the arbitration agreement is in similar terms. This has led to some interesting developments.

For example, *London Borough of Camden* v. *Thomas McInery & Sons Ltd* (1986) presented a new twist and had some interesting things to say about the certification process. The contractors built 219 dwellings for Camden, including a high-rise block with brick cladding. Although Camden's director of architecture was the architect named in the JCT 63 contract, day-to-day supervision was provided by job architects.

The agreed final account was audited and a technical assistant in the quantity surveyor's department prepared interim and final certificates showing the amounts due to McInery. However, when an inspection by the job architect revealed cracks in the brickwork, he attached notes to the certificates saying 'Don't pay'. The certificates had been signed by the chief quantity surveyor in the name of the director of architecture, and there was no dispute about his authority to sign.

Just before the trial of Camden's action, the certificates came to light. McInery argued three points as preliminary issues:

- The signed final certificate was a valid final certificate issued to the employer. It was thus conclusive evidence under JCT 63, clause 30 (7) (now JCT 80, clause 30.9) that the works had been properly carried out and completed. In light of *Crouch* the court had no right to go behind the certificate.
- Even if the certificate was not issued to the employer, it still represented the architect's opinion that the works had been properly carried out and completed. Only an arbitrator, and not the court, could review and revise the architect's expressed opinion.
- Because an interim certificate must show 'the total value of the work properly executed' (JCT 63, clause 30(2), JCT 80, clause 30.2), every interim certificate represented the architect's opinion that the work had been properly executed. Once again only an arbitrator could revise that opinion.

These novel arguments were all rejected by His Honour Judge Esyr Lewis QC, who noted the circumstances in which the disputed certificates were prepared. Those concerned with the certification process had clearly not intended that McInery should receive any further money once the alleged bad workmanship was discovered.

Clause 3(8) of JCT 63 (JCT 80, clause 5.8) requires that 'any certificate ... shall ... be issued to the employer' with a duplicate copy to the contractor. The main issue was whether the certificate was so 'issued'. The judge referred to remarks made by Lord Justice Edmund Davies in *Token Construction Co. Ltd* v. *Charlton Estates Ltd* (1973), who said:

'I have some difficulty in thinking that there would be sufficient compliance with [the contract] if the architect certified in writing and then locked the document away and told no one about it.'

That statement was apposite to the facts.

The mere signing of the certificate was not enough to satisfy clause 3(8) even though the signatory, and the person in whose name it was signed, were in Camden's employment. Judge Lewis was firmly of the view that neither certificate had ever been issued. Did the mere signing of the final certificate amount to the expression of the architect's opinion that the works had been 'properly carried out and completed'? If so, then McInery's argument was sound because the court would have no jurisdiction to determine Camden's claim after the ruling in *Crouch*.

Since the final certificate was not issued, the judge took the view that it was of no effect, and so the *Crouch* principle did not apply. The final certificate, he ruled, was not binding on the parties.

'I can see no reason why an architect should not change his mind after signing a certificate before issuing it. By writing his signature the architect is not irrevocably committed to the opinion which the certificate purports to give ... [The] final certificate only comes to life as a document which is legally enforceable as a certificate, opinion or decision of the architect if he issues it as required by [the] clause.'

The judge also refused to accept McInery's point about the status of the interim certificates. It did not follow that each carried with it the opinion of the architect that the work done up to the date of the certificate had been properly done. The last interim certificate had no force or effect because it had not been issued. Although there was no doubt about the earlier certificates, all of which had been issued and acted upon, 'the primary purpose of the interim certificates is to ensure that the contractor receives regular stage payments,' said the judge.

Clause 30(8) of the contract was very clear in its terms. It says that, except for the final certificate,

'no certificate of the architect shall of itself be conclusive evidence that any work, materials or goods to which it relates are in accordance with the contract.' (JCT 80, clause 30.10.)

Although the architect is duty-bound only to certify payment for work properly done when he issues an interim certificate,

'this does not mean that the architect has expressed a final view as to the state of the works. He is plainly free to take a fresh view each time he issues an interim certificate and his opinion cannot become conclusive until he issues his final certificate.'

In short, the interim certificates were not binding on Camden as to the quality of McInery's workmanship, and Camden's claim did not involve opening up or reviewing the architect's certificates. 'There is in reality no opinion, decision or certificate to open up in the context of this case.'

What the employer must pay

The Court of Appeal's judgments in the case of *Lubenham Fidelities & Investment Co.* v. *South Pembrokeshire District Council and Wigley Fox Partnership* (1986), should be studied by contractors and architects alike. The court said:

> 'We have reached this conclusion with some reluctance, because the negligence of [the architects' practice] was undoubtedly the source from which this unfortunate sequence of events began to flow, but its negligence was overtaken and overwhelmed by the serious breach of contract by (the contractor).'

Two contracts for building work were in JCT 63 terms. The original contractor went into liquidation, and Lubenham, the bondsman, elected to perform the contract itself. It engaged sub-contractors to carry out the work, but remained primarily responsible to the council.

By the end of May 1977 serious difficulties were being encountered on both sides. On 3 June 1977, Wigley Fox, the architects' practice under the contracts, issued two interim certificates and wrongfully deducted sums in respect of both alleged defective work and for liquidated damages to 20 May, although the contractual completion date was 3 September 1977, so it had jumped the gun.

Lubenham protested and said the council's refusal to pay more than the sums shown on the certificates was a repudiatory breach. It also ordered its sub-contractors to withdraw from site. At the end of June 1977 Wigley Fox served a default notice under JCT 63, clause 26(1)(a), alleging lack of regular and diligent progress and a total cessation of the work. Lubenham retaliated by serving notice under clause 25 on grounds of non-payment and thus, the parties were on a collision course.

The trial judge found in favour of the council and against Lubenham and, while he found that Wigley Fox had been in breach of its duty to both the employer and the contractor, he held that the breach of duty was not the direct cause of the damage which it had suffered. The firm of architects was off the hook.

The Court of Appeal, upholding the trial judge and dismissing Lubenham's appeal, made a number of important rulings which are of general application. Of equal importance is the discussion of the circumstances in which the architect can be liable to the contractor. The Court of Appeal held:

- The architect's purported deductions for alleged defective work and liquidated damages were invalid. JCT 63, clause 30(2) (JCT 80, clause 30.2) does not permit such deductions and, of course, the claim for liquidated damages was premature since the contract completion date had not passed;
- the employer's obligation is to pay only the sum stated as due in the certificate, and provided he does so he is not in breach of contract. Where a certificate is erroneous:

  'the proper remedy available to the contractor is ... to request the architect to make the appropriate adjustment in another certificate or, if he declines to do so, to take the dispute to arbitration.'

  This is so whether the error is obvious or a hidden one. The work valued is an approximation only and the application of clause 30 – and similarly worded clauses in other JCT forms – 'to any given set of facts must again involve a degree of assessment by the architect based on personal opinion';
- the issue of an interim certificate is a condition precedent to payment, and the contractors could not recover the amount allegedly due to them by relying on implied terms, such as those recognised in the well-known *Panamena* case (1947). The deductions were not made by the council but, erroneously, by the architect and the court attached great significance to the presence of the wide arbitration clause, which gave the contractor the remedy it needed;
- the council's losses were caused by the contractor's breach of contract, i.e. in withdrawing from the works and refusing to resume after service of the council's clause 26 notice of default;
- Lubenham had no claim against the architect because by persisting in suspension of the works it had broken the chain of causation. Continued suspension by Lubenham after service of the default notice could not, in their lordships' view, have been foreseen nor was it the natural and probable consequence of the disputed certificates;
- for like reasons, the council had no right of recovery against the negligent firm of architects;
- Wigley Fox, architect, was not liable to Lubenham for procuring non-payment by the council of sums due nor for the tort of wrongful interference with the contract because, among other things, there was no intent to interfere with the contract.

By issuing a defective certificate the architect is not thereby interfering with the contract. But the court emphasised that an architect may be so liable if he *deliberately* (rather than incompetently) misapplies the payment provisions with the intention of depriving the contractor of his money.

The case recognised expressly that, since *Sutcliffe* v. *Thackrah* (1974) the architect owes a duty to both employer and contractor to act fairly, just as he does when certifying. Unfortunately, a later Court of Appeal has held that it is not 'just and reasonable' to superimpose remedies beyond those freely agreed in the contractual structure, with the result that, save in the most exceptional cases, the contractor who suffers financial loss because of the architect's negligence or unfairness will have no remedy in tort against the architect: see *Pacific Associates Inc.* v. *Baxter* (1988), discussed on pp. 52–56.

Dealing with the contractor's claim that Wigley Fox had interfered with the contract's operation, the court said it had not.

'On the contrary, albeit in a misguided manner, it was seeking to further the performance of those contracts. As the judge correctly said, "they were doing their incompetent best".'

But, added Lord Justice May, Lubenham's proper remedy, if any, lay in negligence, having earlier refused to accept 'the broad contention that an architect, in effecting an interim valuation under this form of building contract, could never in any circumstances expose himself to a claim' in tort for interference with contract.

Stricter view of certificates

An employer was ordered to pay a contractor £544 602, costs and agreed interest of £89 000 in a case arising out of £2.25 million refurbishment and repair contract carried out under JCT 63 terms between November 1963 and May 1965. The main practical interest of the case – *Rush & Tompkins Ltd* v. *Deaner* (1987) – lies in its discussion of the nature and effect of final certificates.

The case was characterised by the contractor's allegations of withholding of certificates, reluctance and failure to pay sums certified and refusal of access to the site in order to carry out remedial works. The employer riposted with allegations of defective work, failure to complete on time, and claims that the Practical Completion and Final Certificates issued were not properly issued or authorised in accordance with the terms of the contract, but the legal issue fell within a small compass.

An interim certificate for £8349 plus VAT (£1252.35) was issued in the contractor's favour on 27 March 1985. The employer paid only the VAT. After solicitors' correspondence, three further certificates amounting to £151565 plus VAT were issued, against which the employer paid only £40864.

A writ was taken out for £141735.76, practical completion being certified by the architect on 13 May 1986. The architect issued a subsequent schedule of defects and instructed the carrying out of remedial works, but effectively restricting the contractor's access by denying him access for 'administrative purposes, WC or mess facilities or power supply'.

After notice under clause 2(1) the employer employed and paid other contractors to carry out the remedial works on the basis that he (the employer) 'expected the appropriate financial adjustments to be made.'

The Final Certificate was issued in November 1986, and the Contract Sum was adjusted in favour of the employer by an allowance of £42363, which was agreed in the final valuation to represent the contractor's responsibility for incomplete and defective work set out in the architect's snagging list.

The final certificate, which showed deductions in respect of amounts previously certified, related to sums certified and not paid. It also expressly excluded VAT. The employer failed to pay the whole or any part of the sum claimed, and the contractor issued a further writ.

The total amount of all the contractor's claims came to £544602.26. The contractor applied for summary judgments, while the employer raised cross-claims and asked that the dispute go to arbitration on the basis that his claims exceeded, or at least matched, the contractor's claims.

His Honour Judge John Davies QC, official referee, held that the differences between the form of the final certificate and the terms of JCT 63 did not invalidate the certificate. The certificate should have shown the total of the amounts already *paid* to the contractor under interim certificates. The only issue was whether the adjustment of the contract sum by the deduction of £42623 in respect of defective works was in accordance with the terms of the contract.

The judge held that it was. The architect made an adjustment in the Final Certificate. Clause 30(6)(b) provides:

'The Final Certificate shall state ... (b) The Contract Sum adjusted as necessary in accordance with the terms of these Conditions'.

Clause 30(7)(a)(ii) says that the Final Certificate shall be

'conclusive evidence that any necessary effect has been given to all
the terms of this Contract which require an adjustment to be made
to the Contract Sum'.

In the judge's view, that ended the matter.
Judge Davies said:

'The Final Certificate is based on the final measurement and
valuation of all the works. It subsumes all interim certificates. If, and
to the extent, therefore, that the qualification of the balance due
under the final certificate is mistakenly based on sums certified
rather than sums certified and paid, it follows that the sum due
under an unpaid interim certificate must have the same immunity
under clause 35(3) as the balance due under a Final Certificate, and
the employer ought not to be allowed to assert to the contrary when
the imbalance in the Final Certificate has been due to the fault of the
agent.'

The certified amount is not reviewable in arbitration.
   Rush & Tompkins Ltd v. *Deaner* shows a practical approach to the
problems of unpaid certificates and employer's cross-claims. The
employer argued that his cross-claims (which he put at £350 000 for
defective work and £95 000 for liquidated damages) would have to go
to arbitration. If this was so, ruled the judge, the court had no blanket
power to order a stay of execution.
   In reaching this conclusion, Judge Davies referred to *Tubeworkers Ltd*
v. *Tilbury Construction Co.* (1985), a Court of Appeal decision, which
involved a main contractor's claim against a nominated sub-
contractor. The judge took the view that there were no 'special
circumstances which render it inexpedient to enforce the judgment'.
No appeal was made against his decision.

Final certificate?

Clause 30.8 JCT 80 provides for the issue of the final certificate by the
architect. The equivalent provision in the 5th edition of the *ICE
Conditions* is clause 60(3), and in both cases its issue is of importance to
the contractor.
   The architect or engineer has it within his power, of course, to
extend the time for considering the final account by reasonably asking

for more information but, under both contract forms, the duty to issue the certificate is mandatory, once the procedures have been complied with.

The nature and effect of the final certificate is a matter of some significance. It sometimes happens that after its issue, the architect will attempt to argue that what seemed to be a final certificate was in fact not.

In *London Borough of Merton* v. *Lowe* (1981) for example, a 'final certificate' was issued under the 1963 edition of the JCT Form, at the contractor's request. It was expressed to be 'instalment No 28 and final', and, in a covering letter sent by the architect to the contractor, it was described as 'the final certificate'. The Court of Appeal ruled that the certificate and covering letter were to be read together and so it could not be argued that the certificate was not the final certificate.

This can, of course, cut both ways. If, as is usual, a standard printed form of certificate is used, and there is a qualifying covering letter, it may be that the certificate is deprived of its finality. The certificate *must* purport to be final: see the old case of *Brunsdon* v. *Staines Local Board* (1884). It cannot, therefore, be qualified in any way but, as Hudson's *Building Contracts* (10th edition, p. 483) points out, it 'cannot be attacked because it is based upon erroneous reports of an agent of the employer or of the certifier'.

Under the JCT Form, the nature and effect of the final certificate is dealt with by clause 30.9, each word of which is important. It says, that with the exceptions given in clauses 30.9.2 and 30.9.3 and, 'save in respect of fraud', the final certificate shall be conclusive evidence

'that where the quality of the materials or the standard of workmanship are to be to the reasonable satisfaction of the architect the same are to such satisfaction'

and also that

'any necessary effect has been given to all the terms of this contract which require that an amount is to be added to or deducted from the contract sum'

save for accidental inclusions or exclusions or arithmetical errors.

The certificate will be conclusive and binding on both parties even if it is given negligently, although it is clear that a negligent architect is liable to his employer. In *Sutcliffe* v. *Thackrah* (1974), the defendant, who was acting as both architect and quantity surveyor, issued interim certificates. The contractor became insolvent. The employer successfully sued the architect for negligence in certifying defective work for which he had paid. The *Sutcliffe* case applies equally to final certificates.

However, the Court of Appeal's ruling in *Pacific Associates Inc.* v. *Baxter* (1988) apparently debars contractors from suing negligent certifiers in tort for negligence in the issue of certificates, even if they can prove resultant loss.

A further JCT 80 provision should be noted. Clause 30.10 says that:

'save as aforesaid no certificate of the architect shall of itself be conclusive evidence that any works, materials or goods to which it relates are in accordance with this contract'.

If the matter comes before an arbitrator he must accept it as conclusive as to the items stated above. He cannot re-open it save in the exceptional circumstances set out in clause 30.9, i.e. fraud or proceedings commenced before or within 14 days of issue. Another express exception is, of course 'any accidental inclusion or exclusion or ... arithmetical error', and these are not uncommon.

However, as Keating points out (*Building Contracts*, 4th edition) it may be that the point at issue between the parties is not within the range of matters upon which the certificate is stated to be conclusively evidence. He suggests, as examples, that contractors' claims for damages for delay in the issue of drawings and employers' claims for consequential losses are not caught. This last point is not free from doubt, but the author calls in aid the views expressed by Lord Diplock in his dissenting judgment in *Hosier and Dickinson Ltd* v. *P and M Kaye Ltd* (1972), Lord Diplock's view was that the effect of the final certificate:

'was that everything which had to be done by way of building operations had been done and all defects made good but there was no exclusion of claims in respect of alleged past defects and their consequences.'

If this is indeed the law, then the employer is entitled to recover consequential losses, both under the 1963 and the 1980 editions of the JCT Form.

Delay by the architect in issuing the final certificate does not prevent it being conclusive when issued. However, some serious irregularity in its issue could affect the position, for then it would not be the certificate required by the contract at all. The certificate must be given by the person authorised by the contract, i.e., the named architect (or his duly appointed successor) under the JCT Form.

The certifier may also be disqualified from acting in that capacity, but this would be exceptional, and the case law is sparse. For example, in *Kimberley* v. *Dick* (1871), a secret arrangement between the architect and employer to limit the cost to a certain sum was held to disqualify, and in *Kemp* v. *Rose* (1858), a private arrangement which had the effect of inducing the architect unfaily to cut down the cost of the works was held to have a similar effect. The common factor is that the matter alleged to disqualify must be such as to appear to affect his impartiality.

The duty to be impartial is vital. In issuing the final, or any certificate, the architect must act impartially as between the parties.

'If the architect has failed to preserve that attitude of independence required of him in the discharge of his responsible and ... difficult duties, as for example if he has allowed the (employer) to influence him in the mistaken view that the architect's duty is to observe the (employer's) wishes, then, the architect becomes disqualified' Emden's *Building Contracts*, (8th edition, vol 1, p. 112).

Such instances will be rare today.

Suspending work for late payment

Late payment of certificates by building employers seems to be on the increase and many contractors face cash flow problems because of it. The standard contracts are quite specific in their terms. Under JCT 80, for example, the contractor is entitled to be paid within 14 days from the *issue* of the certificate by the architect, and the employer is only entitled to deduct certain specific and liquidated sums, e.g. liquidated and ascertained damages.

The contractor's remedies for late or no payment are, however, very limited. His principal solution is to sue on the certificate and ask for summary judgment under Order 14 of the Rules of the Supreme Court and he will, in general, get it, although it is possible that the employer will be able to raise a *bona fide* arguable contention that the certificate has been overvalued and in that case he is entitled to have the issue arbitrated and the contractor will not be entitled to summary judgment: *G. M. Pillings & Co. Ltd* v. *Kent Investments Ltd* (1985). He certainly cannot suspend work, since he has no contractual right to do so under JCT 80, in contrast to the unpaid nominated sub-contractor operating under Form NSC4 or 4a, where clause 21.8 confers that right upon him. However, a *virtual* suspension of work on a contract in JCT 63 terms, where the contractor was held by the Court of Appeal not to amount to an unlawful repudiation of the contract in the well-known case of *J. M. Hill & Sons Ltd* v. *London Borough of Camden* (1981).

The contractor's real remedy is to determine his own employment under JCT 80, clause 28.1.1, which lists as a ground for determination the employer's failure to pay the amount properly due to the contractor on any certificate within 14 days from its issue. The contractor must then issue a notice stating that he will determine his own employment unless payment is made within seven days of receipt of the notice, which should be served by registered or recorded delivery post. In practice this is very much the last resort and I am afraid that many large employers take advantage of the fact that few contractors are likely to go to such extremes.

The problem is not limited to the UK and one New Zealand case gives a ray of hope. It is *Fernbrook Trading Co. Ltd* v. *Taggart* (1979) where there was a roadworks contract and, among other things, the employer was in breach of contract by failing to make progress payments for a total of 23 weeks and this caused the contractor financial difficulties and led to his reducing labour on the site. The engineer was also at fault; he failed to issue progress payment certificates for five separate months.

The contract overran, and the question arose as to the contractor's liability for liquidated and ascertained damages amounting to $2500. The Supreme Court of New Zealand, in a very long and careful judgment, held that the engineer had no power to extend time because of the employer's breach of contract in failing to make progress payments, and further that the breach set the completion date at large with the effect that Taggart was not liable to pay liquidated damages. The position would in all probability be the same under the JCT contracts. It would be different under ACA2 because whichever alternative of clause 11.5 is used, the architect is empowered to extend time for delay caused by:

'any act, instruction, default or omission of the employer, or of the architect on his behalf, whether authorised by or in breach of this Agreement.'

Luckily, the courts take a fairly robust approach when exercising jurisdiction under Order 14 procedures, and tardy employers can hardly complain if they are sued to judgment when a certificate is not paid. The procedure is fairly speedy and the approach of the courts to the problem is neatly illustrated by *Ellis Mechanical Services Ltd* v. *Wates Construction Ltd* (1976), where there was £52 437 retention money being held in respect of a nominated sub-contractor's work, and this formed part of the plaintiff's total claim of £140 665. They sued for the full sum and sought summary judgment under Order 14. The Court of Appeal ruled that the plaintiffs should have judgment for the £52 437 retention money, with interest at 9% from the date of the writ, while the balance was referred to arbitration.

Lord Justice Lawton indicated the right approach:

> 'The courts are aware of what happens in these building disputes; cases go either to arbitration or before an official referee; they drag on and on; the cash flow is held up. In the majority of cases, because one party or the other cannot wait any longer for the money, there is some kind of compromise, very often not based on the justice of the case but on the financial situation of one of the parties. That sort of result is to be avoided if possible. In my judgment it can be avoided if the courts make a robust approach, as the Master did in this case, to the jurisdiction under Order 14.'

These are strong words, and seem to have had an effect, and certainly the judges do not like employer's attempts to postpone the evil day of payment.

Contractors are not naturally litigious, but if certificates are persistently paid late or not at all they must take steps to protect their interests, by suing on the certificate if necessary. For my part, I would like to see the main contract forms conferring on the contractor a right to suspend work for nonpayment, following the NSC4 or NSC4a pattern. Indeed, since architects often fail to issue certificates at the right time – if my contractor friends are to be believed – there should also be power to suspend work if the architect fails to issue any certificate at the date required by the contract. Such a power is contained in clause 33(4) of the Singapore Institute of Architects' Standard Form of Contract which adds that:

> 'the cost of such suspension shall be borne by the employer and the contractor shall be entitled to an extension of time therefore ...'

Perhaps it is too much to hope that the Joint Contracts Tribunal will grasp this nettle.

Chapter 5

Liquidated Damages and Extensions of Time

When time is not always money

Assessing a contractor's entitlement to an extension of time is a
difficult task. It is not an exact science and the length of any extension
can seldom be calculated down to the last day or fraction of a day. The
JCT forms make this clear by their use of the word 'estimate', but
contractors often take the contrary view on the mistaken assumption
that a successful money claim depends on the determination of the
extension of time.

This heresy was finally extinguished by the case of *H. Fairweather* v.
London Borough of Wandsworth (1988), which was an appeal on questions
of law from an arbitrator's interim award. The arbitrator found that
where an event occurs on site which causes delay to completion and
which can be ascribed to more than one of the reasons specified in JCT
63, clause 23, as there was no mechanism in the contract for allocating
an extension between different heads, the extension must be granted
in respect of the 'dominant reason'. Since the dominant reason was a
series of strikes (for which an 81-week extension was granted) which
carried no entitlement to loss and/or expense, the contractor was not
entitled to his prolongation costs.

The contractor wished to have 18 of the 81 weeks reallocated
against late information and architect's instructions. This part of the
award was remitted to the arbitrator for reconsideration, because it
was incorrect to allocate extensions of time according to the dominant
cause of the delay. The grant of an extension of time is not a condition
precedent to reimbursement of direct loss and/or expense under JCT
terms.

This raises some practical problems, which were clearly in the mind
of His Honour Judge James Fox-Andrews QC. He made the important
point that ordinarily one practical effects:

'will be that if the architect has refused an extension of time (under
clause 23(f) for late information) the contractor is unlikely to be

successful with the architect on an application under condition 24(1)(a).'

This, of course, applies to claims under analogous contracts as well where the ground is employer responsibility.

Judge Fox-Andrews gave an example to illustrate the point. A contract is entered into in May with completion on 31 July in the following year. No work can be carried out from 1 November to 31 March for seasonal reasons and during that time the contractor's plant is lying idle. This is known to the parties when they make the contract. In early April the works are on programme; the architect issues a variation order (VO) which will add three months to the contract period and he grants an extension to 31 October. In mid-October, when the works are still on course, there is a strike which lasts until 31 March. The contractor restarts work on 1 April.

Because he has had no opportunity to protect his machinery during the strike period, it takes the contractor a further two months to complete.

The architect grants an eight-month extension of time because of the labour troubles. The judge said that there was no reason why, under JCT 63, the contractor could still not recover all his direct loss or expense under clause 11(6).

This approach is practical, but it is very difficult to see how the extra cost which resulted from the strike can be said to be loss or expense flowing *directly* from the variation order, because the principle of causation requires the loss to have been *caused* by the variation order and not merely be the occasion of it. The architect or quantity surveyor must be satisfied that the contractor has incurred loss or expense as a direct result of the VO.

Although JCT 63 contains no mechanism for allocating extensions under different heads, JCT 80 clause 26.3 does so since it provides that if it is necessary for the purpose of ascertainment of loss and/or expense, the architect is to state in writing to the contractor what extension of time he has granted under clause 25 in respect of those events which are also grounds for reimbursement under clause 26. That would not, of course, assist in the judge's example because the whole of the architect's extension was in respect of the strike.

The purpose of the extension of time and money claims provisions is entirely different, as *Fairweather* v. *Wandsworth Borough Council* emphasises. The practical problems arise when an extension is granted because of neutral events. If an extension is granted for a matter which is the employer's responsibility – late information or a variation order, for example – it follows almost inevitably that the contractor will be able to recover his directly incurred costs. But of course to do this he must show that he has not been adequately reimbursed under some other contractual provision, such as the variation clause itself, and he can establish that the direct loss and/or expense flows from the event relied on.

It is now plain that the 'dominant' test is wrong – though it has been widely adopted in practice.

The extension of time clause stands on its own. It does not trigger off and condition the right of the contractor to recover loss and expense where applicable.

As is usual in cases of this kind, the parties had come to a compromise agreement at the end of the contract, and the judge had to consider whether this covered both the amount of the loss and expense recoverable in respect of a 34-week period *and* the amount of the period of extension of time granted. Fairweather's quantity surveyors had confirmed their 'full and final agreement to the figures' contained in the employer's quantity surveyors' letter. The contractor had been granted the 34-week period on various grounds. He then applied for ascertainment of loss and expense. 'An application in these circumstances … in itself is not an acceptance of the extension of time allowed', said the judge. On the evidence he decided that there was a binding agreement only as to the amount of money due in respect of the period of time which the architect had granted and it was not an agreement as to the period granted.

Notice of extension

The question of whether the contractor's written notice is a condition precedent to the architect's grant of an extension of time under JCT 63 terms was, rather surprisingly, one of the bones of contention in *London Borough of Merton* v. *Stanley Hugh Leach Ltd* (1985). The council argued that the contractor's notice under what is now JCT 80, clause 25.2.1.1. was to be treated in this way.

That argument was based on the assumption that the notice is necessary to warn the architect about the existence of an event causing delay to the contract, and of which the architect would otherwise be ignorant, and so not in a position to take remedial action to diminish the effect of the delay. This argument, of course, assumes that the architect has little or no knowledge of what is happening on site, and that seems to do a disservice to conscientious architects.

By the terms of his appointment, the architect must make periodic visits to the site to check progress and to determine whether the contractor is proceeding in accordance with the contract documents. The contract imposes on the contractor a duty to proceed 'regularly and diligently' with the works, and so this is certainly one of the matters to which the architect must direct his mind.

Merton's case was that the architect is under no duty to consider or form an opinion on the question whether completion of the works is likely to be delayed unless, and until, the contractor has given notice of the cause of the delay that has become 'reasonably apparent' to him.

The High Court's analysis of this argument is an interesting one, and took the form of a comparison between the circumstances in which a contractor is required to give notice on the one hand, and those in which the architect is required to form an opinion, on the other. The first part of JCT 63, clause 23, looked to the situation where it is apparent to the contractor that the progress of the works is delayed – the wording has changed in JCT 80 so as to cover actual and likely delays. Under JCT 63 there must be an event known to the contractor which has resulted or will inevitably result in delay; under JCT 80 he must also do this when progress of the works is likely to be delayed in the future.

The second part of the clause looks at a situation in which the architect has formed an opinion that completion is likely to be, or has been delayed. Obviously the architect might know of some delaying events likely to cause delay to completion but which have not caused an actual or prospective delay in the progress of the work which is apparent to the contractor.

If he forms the opinion that progress of the works is likely to be delayed beyond the current date, he must estimate that delay and grant an appropriate extension. 'He owes that duty not only to the contractor but also to the building owner,' said Mr Justice Vinelott.

Grammar aside, Merton's contention led to a consequence that was clearly not intended, namely that the architect can ignore events which he knows are likely to cause delay beyond the completion date, even though to his knowledge, the contractor is not aware that the progress of the works is delayed.

The judge was not in favour of that contention, and he felt that a rational result could be achieved by reading the word 'and' in clause 23 (JCT 80, clause 25.3.1.1.) as joining 'two related but independent duties.'

Failure by the contractor to give notice of a cause of a delay when it *is* reasonably apparent to him that 'progress of the works is being or is likely to be delayed' will, therefore, leave his right to an extension of time unaffected. The contractor's failure to give notice is a breach of contract, and the judge expressly ruled that if no notice is given by the contractor, or he gives it late, and in consequence to the architect does not become aware that completion is likely to be delayed as a result, the architect can take this factor into account in assessing the extension of time grantable.

In the result, since the contractor cannot benefit from his own breach, he cannot receive a greater extension than he would have received, had he given notice at the proper time when the architect might have been able to avoid, or reduce the delay by some valid instruction or requirement.

This is clearly correct, and even before *Merton* v. *Leach* it was the generally accepted view in the construction industry that the contractor's notice was not a condition precedent to the operation of the extension of time provisions.

The terms of the High Court judgment clearly support the view that the contractor is under a general duty to notify *any* delay to progress: it is not a duty limited to notifying the architect of delays caused by the specified events which give rise to an extension of time.

Indeed, this is inherent in the clause, and any other interpretation would make nonsense of it. Architects who wrongly assume that the contractor's notice is a condition precedent – as many did before *London Borough of Merton* v. *Leach* – are acting against the employer's interest if they refuse to consider a cause of delay of which late notice is given, or of which they have knowledge, but no notice from the contractor. That course means that the architect is refusing to perform his duties, and the employer forfeits his right to liquidated and ascertained damages – and the architect is likely to be the unhappy recipient of a writ for professional negligence.

Where the contractor gives notice of delay, the architect is of course entitled to issue an instruction (if empowered by the contract) aimed at reducing the overall delay, which becomes the contractor's entitlement to an extension of time if that delay causes the contract to overrun.

This he might do, for example, by omitting part of the works. If he does not do so, the contractor remains entitled to an extension equal to the overall delay based on the fact that no such instruction was actually issued.

The principles behind this part of the ruling in *Merton* v. *Leach* apply not only to JCT 63 contracts, but to JCT 80, and to the JCT Intermediate Form IFC 84 as well. No standard form construction contract makes a notice or application by the contractor a precondition to the grant of an extension of time, although some people suggest that Minor Works 80 does so. Such an interpretation would make nonsense of the provisions and put the employer's interests at risk, since failure properly to grant an extension of time for delay for which the employer is responsible in law would invalidate the provision for liquidated and ascertained damages.

Delay and damages

Contractors often complain that architects and engineers are parsimonious in the granting of extensions of time. That would certainly seem to have been the intention of the drafstman of JCT 80 which introduced the very valuable architect's review of extensions of time in clause 25.3.3. The architect must carry out the review in the light of any causes of delay – called somewhat inelegantly 'relevant events' – whether or not they have been notified to him by the contractor.

The object of this, of course, is to preserve the employer's right to liquidated damages. If the architect fails to carry out his review and grant any appropriate extension of time, then the contract time can become 'at large'. There is then no date from which liquidated damages can run, and hence none will be recoverable. The employer would be left to claim general damages at common law – on the basis of his proven losses. If time is 'at large' the contract completion date ceases to be applicable and the contractor's obligation becomes one to complete 'within a reasonable time'.

What is a reasonable time will depend on all the facts and circumstances. As a rough guide, it is probably the original contract period plus a period equivalent to any extension that ought reasonably to have been granted had the architect acted correctly. Despite the mandatory wording of clause 25.3.3., Lord Justice Croom-Johnson suggested that it is 'directory only as to time and is not something which would invalidate the [recovery] of liquidated damages' in *Temloc Ltd* v. *Errill Properties Ltd* (1987), but why this should be so is not clear and the contract wording and intention is plain.

The problem was more acute for those working under JCT 63 contracts because of its totally defective extension of time clause. JCT 63 did not contain a reviewing power, and also lacks a number of grounds now found in JCT 80, clause 25, e.g. 'the supply by the employer of materials and goods which the employer has agreed to provide for the works or the failure' so to supply.

Moreover, the architect operating under JCT 63 probably had no power to grant an extension of time once the due date for completion has passed. This was of great importance in respect of events giving rise to an extension of time which are the employer's fault or responsibility, such as late information and so on.

It is quite clear and settled law – despite assertions to the contrary –
that if the employer is wholly or partly responsible for the delay or if
an extension is not made at the right time, then the liquidated damages
clause becomes inoperative and time becomes at large: *Peak Construction
(Liverpool) Ltd* v. *McKinney Foundations Ltd* (1970). The holding was quite
clear:

'If the employer is in any way responsible for the failure to achieve
the completion date, he can recover no liquidated damages at all, and
is left to prove such general damages as he may have suffered.'

In *Rapid Building Group* v. *Ealing Family Housing Association Ltd* (1984) the
Court of Appeal upheld the *Peak* case as binding authority. Lord Justice
Stephenson said:

'... if the employer is responsible for any delay which does not fall
within the *de minimis* rule, it cannot be reasonable for him to have the
works completed on the completion date. Whatever the reason
underlying the decision of this court it binds us ...'

The *Rapid Building* case left open the point as to whether if the
employer's claim to liquidated damages has disappeared, his claim for
unliquidated damages is subject to a ceiling equal to the original but
defunct liquidated damages provision. This remains to be decided, but
it is a sustainable view that any claim for general damages has on it a
ceiling equal to the amount of the failed liquidated damages provision.
Lord Justice Lloyd also threw out a tantalising point:

'Like Lord Justice Phillimore in (the *Peak* case) I was somewhat
startled to be told in the course of argument that if any part of the
delay was caused by the employer, no matter how slight, then the
liquidated damages clause in the contract becomes inoperative. I can
well understand how that must necessarily be so in a case where the
delay is indivisible and there is dispute as to the extent of the
employer's responsibility for that delay.
   But where there are, as it were, two separate and distinct periods
of delay with two separate causes, and where the dispute relates
only to one of those two causes, then it would seem to me just and
convenient that the employer should be able to claim liquidated
damages in relation to the other period.'

This statement was not part of the holding in *Rapid Building* but, coming as it does from a Lord Justice of Appeal must carry a good deal of weight and sooner or later it will be seized upon and argued out.

In principle, once the employer's right to liquidated damages has gone it cannot be revived. He is left to pursue his other remedy and claim unliquidated damages, probably subject to a ceiling on its amount. Presumably what Lord Justice Lloyd had in mind was the situation where there is a neutral event which is the fault of neither party, such as exceptionally inclement weather, and one which is the employer's fault.

Even in that case, with respect, it is hard to accept that it is 'just and convenient' to allow the employer part liquidated damages. But perhaps his lordship had something else in mind, e.g. where the contractor is himself in default and then there is delay caused by the employer.

However, even then the sensible solution would be for the employer to pursue a damages claim at common law, even if he is then put to proof of loss. For the moment, the position is as stated in *Peak*. If completion by the specified date is prevented, wholly or partly, by the fault of the employer, he can recover no liquidated damages unless there is an extension of time clause providing for an extension of time on that ground, and the architect or engineer grants an extension of time as specified in the contract: *Percy Bilton Ltd* v. *Greater London Council* (1982), a decision of the House of Lords.

Time at large

When a contract runs into delay, contractors often claim that time has become 'at large', i.e. that the contract completion date is no longer applicable, and the obligation is then to complete within 'a reasonable time'. This argument was fairly easy to run successfully under JCT 63 terms where delays were caused by the employer (or his architect) and clause 23 made no provision for an extension of time for this cause.

Perhaps the most notorious example is *Peak Construction (Liverpool) Ltd* v. *McKinney Foundations Ltd* (1970), where an in-house construction contract empowered the architect to grant extensions of time for, *inter alia*, extras and additions, *force majeure* and 'other unavoidable circumstances'. The Liverpool Corporation (whose contract it was) was held by the Court of Appeal as not entitled to recover any liquidated damages against the plaintiffs because they were to blame for a great part of the 58-week delay. They had failed to take action on a consultants' report about defective piling and to authorise necessary remedial works. This was not an 'unavoidable circumstance'. As Lord Justice Salmon put it:

> 'I cannot see how, in the ordinary course, the employer can insist on compliance with a condition if it is partly his own fault that it cannot be fulfilled'.

If the contractor can establish that time has become at large, because the delay is partly the employer's fault and is not covered by the extension of time clause, this can be very much to his advantage, as the liquidated and ascertained damages will cease to be enforceable.

However, contrary to the generalised observations of Lord Justice Salmon in *Peak* to the effect that liquidated damages and extensions of time clauses are *exclusively* for the employer's benefit, such provisions benefit *both* parties. The position is clearly stated in Hudson's *Building Contracts*, 10th edition, p. 624:

> 'An extension of time clause, which *prima facie* would appear to be inserted for the benefit of the contractor, might also be regarded as being for the benefit of the employer since in cases of prevention or breach its function might be to keep alive a liquidated damages clause which would otherwise have been treated as no longer applicable'.

The terms of JCT 80 mean that the 'time at large' argument is now unlikely to succeed in the vast majority of cases because of the new grounds for extension of time which have been added:

- Execution of work not forming part of the contract by the employer's licensees: clause 25.4.8.1.
- Supply or non-supply of goods or materials which the employer has undertaken to supply: clause 25.4.8.2.
- Failure by the employer to give ingress to or egress from the site overland, etc, in his possession and control, as specified in the contract Bills: clause 25.4.12.
- Deferment by the employer of the giving of possession of the site where clause 23.1.2 is stated to apply: clause 25.4.13 (added in 1987).

IFC 84 is in similar terms, and so in neither case is the 'time at large' argument likely to succeed unless the contractor can show that there were delays caused by the employer which are not specifically covered by the extension of time clause. Indeed, even in those cases it will only be to the contractor's advantage to claim that time is at large where there is a very high rate of liquidated and ascertained damages written into the contract. If he establishes that time is at large and still fails to complete in a reasonable time, this merely means that the employer will be put to proof of the damage which he actually suffers.

An important point is that a judge or arbitrator will have to give a strict interpretation to any extension of time clause and that is why clause 11.5 of the ACA form, in either alternative, is probably the best existing extension of time clause. Clause 11.5 enables the architect to extend time for delay caused by *any* act, instruction, default or omission of the employer, or of the architect on his behalf, whether authorised by or in breach of this Agreement.

This is a catch-all provision beneficial to employer and contractor alike. The strict approach to extension of time clauses is shown by the old and oft-quoted case of *Wells* v. *Army & Navy Co-operative Society Ltd* (1902), where a building contract provided for the grant of extensions of time for 'other causes beyond the contractor's control'. The Court ruled that these words did *not* cover delays caused by the interference of the employers or their architect, and no liquidated damages were recoverable. The provisions of ACA, clause 11.5 are, to my mind, different to those in *Wells* and are effective.

Disputes often arise as to the extension of time which is actually granted. This is not an exact science, and all the contracts use the word 'estimate'. Opinions as to the length of an extension of time can vary as between different architects and engineers, and I think that there is quite a wide margin. JCT 80 (and the ICE Civil Engineering Conditions) in fact encourage the architect or engineer to be parsimonious in his initial grant of an extension, because the defect can be cured later when he reviews the position and can then rectify an earlier under-grant. Clause 44 of the ICE Conditions of Contract is, to my mind, vastly superior to its JCT equivalent, especially in the provision of clause 44(4) which says that 'No such final review of the circumstances shall result in a decrease in any extension of time already granted by the Engineer ...'

This is an area where emotions often run high, but the basic principles involved are simple. As usual, the answer is to read the contract carefully and ensure that its provisions are operated correctly.

Adverse weather conditions

'Exceptionally adverse weather conditions' is one of the grounds for extension of time in JCT 80, clause 25.4.2, and IFC 84, clause 2.4.2.

The wording makes it clear that the definition covers extremes of heat and dryness, as well as the more normal English weather. (JCT 63 referred to 'exceptionally *inclement* weather' which would not, it is thought, cover drought-like conditions, even though these might have a serious effect on progress.)

The Government contract – GC/Works/1 Edition 2 – is more precise. It refers (clause 28(2)(b) to 'weather conditions which make continuance of work *impracticable*', while the 5th edition of ICE Conditions is in terms similar to JCT 80. It refers to 'exceptional adverse weather conditions'.

In all cases, however, quite unusual severity is required because the contractor is expected to programme his work to take account of the normal weather conditions to be expected at that time of year in the locality.

Frost is not exceptional in this country in winter months, nor is heavy rain in the spring or autumn. The weather may be exceptional in its duration or intensity but, for it to be a ground for extension of time, it must affect progress of the works.

If it does not, then no extension of time can be claimed or granted. Even if the weather is exceptionally adverse for the time of year there is no right to an extension of time if it does not interfere with the works at the particular stage they have reached. The effect of the adverse weather is to be assessed at the time when the works are actually carried out and not when they were programmed to be carried out, even if the contractor is in culpable delay during the original or extended contract period: *Walter Lawrence & Son Ltd* v. *Commercial Union Properties (UK) Ltd* (1985).

Persistent and torrential rain in London is probably exceptional during the summer months, but not every contractor working there during that period would be entitled to an extension of time. If he was working on the foundations then, I think, in principle he would be so entitled, but not if he was plastering internal partitioning!

How does one determine that weather conditions are 'exceptional'? All the commentators agree that this is to be judged from public records or weather charts covering a long period and relevant to the particular area where the work is to be carried out.

The key word is *exceptional* which refers to the weather itself. The matter is fairly put in the Aqua Group's *Contract Administration*, 5th edition, 1981, p. 68:

'The word "exceptionally" is clearly the important one ... and it must be considered according to the time of year and the conditions envisaged in the contract documents.

Thus, if it were known at the time that a contract was let that the

work was to be carried out during the winter months, and if that
work is delayed by a fortnight of snow and frost during January,
such a delay could not be regarded as due to exceptionally adverse
weather. If, however, such work was held up by a continuous period
of snow and frost, from January until the end of March, an
extension of time would clearly be justified under this clause.'

The location of the work is important, as the authors point out, and
they add that 'in certain circumstances such delays may be avoided by
providing *in the contract* for additional protection and even temporary
heating arrangements.'

Under JCT terms, of course, the contractor is under an obligation to
take all practicable steps to avoid or reduce delay and also to do all that
may reasonably be required to the satisfaction of the architect to
proceed with the works, and this proviso qualifies the right to
extension of time.

Some architects have argued that, as a result, the contractor can be
required to expend substantial sums of money to meet the obligation.
They are fortified by the views expressed by John Parris, *The Standard
Form of Building Contract: JCT 80*, 3rd edition, s. 7.06, but for reasons I
have explained elsewhere (*Building Contract Claims*, 2nd edition, p. 42)
this is not so and hence, presumably, the Aqua Group's reference to
making provision in the contract for temporary works, etc. The
contractor need take only ordinary measures to proceed or, as I have
put it elsewhere, his obligation is 'to show willing.'

Generally, of course, there are concurrent or parallel causes of delay
some of which, like exceptionally adverse weather, merely give rise to
an extension of time and others, like late information, which may give
rise also to a financial claim. In such circumstances the contractor often
argues for an extension of time on the ground which also gives rise –
independently – to a loss and/or expense claim and is not at all happy
if the architect opts for adverse weather as the ground for extension.

This attitude displays a fundamental misconception about the
contractual position. Disruption costs arising from late information
are clearly reimbursable either under the express contract terms or,
alternatively, at common law.

Where there are parallel causes such as those I have mentioned, the
question for the architect is simply: 'Is the progress of the works
materially affected by the late information?' If the answer to that
question is 'no' because no work could have proceeded even had the
information been available on time, i.e. because of the adverse
weather, there is no claim for loss and/or expense.

I accept that this view is not popular with contractors, but it seems to me (and to others) that this is what the contract says. There is no case law on the topic – though it derives some support from observations of Mr Justice Megarry in *Hounslow Borough Council* v. *Twickenham Garden Developments Ltd* (1971), as well as the major textbooks.

At the very best, it is a grey area. Under JCT extension of time clauses it is the effect of the relevant event upon the completion date which is critical. In contrast, under JCT 80 clause 26 it the effect of the event upon *progress* which matters. Thus, if late information has no effect upon the progress of the works (which does not necessarily delay the completion date) there can be no entitlement to loss and/or expense.

Expressing liquidated damages

Liquidated damages in the context of local authority housing refurbishment schemes give rise to many practical problems. It is all too easy for the local authority employer to fall into the trap illustrated by *Bramall & Ogden Ltd* v. *Sheffield City Council* (1983).

In that case, a large housing contract was let on JCT 63 terms. The contract provided for liquidated damages under clause 22 for the contractor's failure to complete 'the Works' by the date for completion. The Appendix entry expressed the rate for liquidated damages as 'at the rate of £20 per week for each uncompleted dwelling'. The contract made no provision for sectional completion, but clause 16(e) provided for a proportionate reduction in the rate of liquidated damages if the employer took possession of part of the Works before completion of the whole. In fact, as the houses were completed, they were taken over in groups.

The employer then sought to claim liquidated damages for some of the uncompleted dwellings at the rate stated in the Appendix. While the parties' intention was reasonably plain, His Honour Judge Lewis Hawser QC held that on a strict interpretation of the contract liquidated damages could not be claimed at all in these circumstances. The Appendix applied to the whole of the Works and not to just some of the houses. Clause 16 could not then operate because of the way in which the rate of liquidated damages was expressed in the Appendix.

The real kernel of the decision is that in the contract both clauses 22 and 16 referred to the whole Works, but the liquidated damages themselves were expressed by reference to a single uncompleted dwelling.

This, of course, raises problems for employers and their advisers, especially in refurbishment contracts, because if the liquidated damages provision is at all uncertain, it will fail and the clause will not bite. The JCT contracts make one attempt to deal with the problem by providing a mechanism for the proportionate reduction of liquidated damages, but it is in practice very difficult to apply and, when work is inevitably divided into phases or smaller units – as with public authority refurbishment contracts – a simpler and more certain way of apportionment is essential. The *Bramall & Ogden Ltd* method is a potential snare and sectional completion is one way out of the difficulty. If the method '£x per unit per week' is used, the unit referred to must be the smallest unit which will ever be handed over.

Another important point is that in such cases you must have a defined starting point – which may cause problems where some of the units are occupied. The usual and most sensible method is to have a detailed handover programme binding on both employer and contractor. In some cases the work content or the period of performance may start as contingent. Even here there must be some certainty, e.g. approximate quantities of work and the outline length of time; the starting time can then be left to run from an order to commence. It is quite clear that in many cases there is a danger that a liquidated damages clause can fail, since it is arguable on the basis of *Peak Construction Ltd* v. *McKinney (Foundations) Ltd* (1970) that in case of any ambiguity at all, the courts will apply the *contra proferentem* rule of interpretation and so interpret any ambiguity against the employer who is seeking to enforce the provision. There is debate as to whether the liquidated damages clauses in negotiated contracts (such as the JCT forms) will be construed *contra proferentem* but the danger is there. Liquidated damages cannot be enforced if the applicable time period cannot be mathematically computed.

If a liquidated damages clause does fail, it is still generally thought that the employer is left with his claim for general damages at common law. Here, the situation is legally in a state of flux. At one time it was thought that where the liquidated damages clause had become defunct, and the employer went for general damages, he might have got an unexpected bonus. Now, the current and sustainable view is that the employer's general damages in this situation would be limited to the amount of the dead liquidated damages clause. This trend certainly seems to follow from such cases as *Temloc Ltd* v. *Errill Properties Ltd* (1987) and *Rapid Building Group Ltd* v. *Ealing Family Housing Association Ltd* (1984). The point has *not* been directly decided, but *Temloc* did decide that liquidated damages are exhaustive of the employer's remedies for the breach of late completion even where, as in that case, the liquidated damages had been expressed as '£NIL'.

It is certain that in today's economic climate any major contractor against whom liquidated damages are claimed may well seek to challenge their validity. Quite apart from the special case of refurbishment contracts, in all cases liquidated damages will cease to be enforceable if part of the delay was caused by the employer (or his architect) and the extension of time clause either does not cover the situation, or has not been properly operated.

This is a very common situation; the JCT contracts, for example, do not cover all possible 'employer defaults' in the extension of time provisions. JCT 80 is a vast improvement on the defunct JCT 63 in this respect, as are the other JCT contracts (except for Minor Works 80). A liquidated damages clause will certainly fail if extensions are not granted for any cause of delay which is a default by or on behalf of the employer, and since architects have no general dispensing power, the default must be clearly covered by the extension of time clause: *Peak Construction (Liverpool) Ltd* v. *McKinney Foundations Ltd* (1970). It may be that the time has come for the industry to rethink it position on liquidated damages. In most cases, the employer's interest is not best served by omitting liquidated damages altogether and relying on his right to general damages at common law, because of all the difficulties involved in proving loss. One possible solution, of course, would be to provide a bonus for early or prompt completion.

A total remedy

The confusion about liquidated damages is shown by the Court of Appeal's decision in *Temloc Ltd* v. *Errill Properties Ltd* (1987), where the contract was in JCT 80 form.

Temloc, a builder, contracted with Errill to construct a large shopping development at Plympton, Devon, for £840 000. The contract was signed in March 1984, and the Appendix entry for liquidated and ascertained damages had been filled in as '£NIL'. The Completion Date was 28 September 1984, but practical completion did not take place until 14 November 1984. Disputes arose between the parties and Temloc issued writs against Errill claiming sums due on the architect's certificates. There were also claims and counterclaims between Errill and businesses which had agreed to occupy parts of the Plympton complex, arising out of the delay in completion. Potentially, large sums of money were at stake.

The preliminary issue before the Court was on the interpretation of clause 24 of JCT 80, read in conjunction with the Appendix entry. The contractor argued that if one read what was in the Appendix, it meant that the liquidated damages were simply to be nil. The employer was not entitled to general damages for late completion. Errill, the developer, took the opposing line, and said that by writing in '£NIL' the parties indicated that clause 24 was in effect to be excluded from the contract altogether, leaving them with their right to claim unliquidated damages at common law for late completion. Both the trial judge and the Court of Appeal concluded that the proper interpretation was not, as Errill contended, that the clause could be discarded altogether from the contract, but was to be left in and expressed to apply in a negative way.

There was some evidence of a course of dealing between the parties, since Temloc and Errill had entered into at least four contracts before. Some of them were in JCT 63 terms, but nothing turned on the difference.

In each of those contracts the Appendix was completed in the same way with the inclusion of the word NIL. The trial judge concluded, on the evidence, that 'this was deliberate as it was agreed that there should be no damages for delay in completion.'

He accepted that as it was not practical to provide a bonus for early completion, there would be no liquidated damages if the contractor was late.

Lord Justice Croom-Johnson said that even if there had been no evidence of a course of dealing, he would have interpreted the contract in exactly the same way.

Errill had no claim for general damages for failure to complete within a reasonable time. On appeal, it was argued that Errill had a choice as to which to go for, whether for liquidated damages or for damages at large. The Court of Appeal adjudged that:

'On the wording of clause 24 there is no choice available. Any such claim for damages at large would have to be based on an implied term in the contract. If clause 24 had been excluded from the contract altogether, as submitted by [Errill], it would have been necessary to imply such a term and give effect to it. But as clause 24 is tied to dates certified by the architect and a method of calculation is provided in the Appendix, there is no room for implying such a term. Clause 24 lays down an agreed provision for calculating (damages for non-completion) by liquidated damages, which is covering all the damages for non-completion'.

In other words, the entry '£NIL' did not mean that there was no effective clause for the recovery of liquidated damages. There was, but the parties had agreed to use it in a negative way.

A further interesting point was that the architect involved had misapplied the provisions of clause 25 of JCT 80 which provided that the final review of extensions of time must be given not later than the expiry of 12 weeks from the date of practical completion. Here that period of 12 weeks was exceeded. The Court of Appeal rejected Errill's contention that the operation of clause 25.3.3 was a precondition to the operation of clause 24.2 which had not been complied with.

As Lord Justice Croom-Johnson put it:

> 'Even if the provision in clause 25.3.3 is applicable, it is directory only as to time and is not something which would invalidate the calculation and payment of liquidated damages. The whole right of recovery of liquidated damages does not depend on whether the architect, over whom the contractor has no control, has given his certificate by the stipulated day'.

This will be of comfort to those many architects who do fail to operate clause 25 correctly, but the last word has not necessarily been said on this point.

If the cause of the delay was wholly or partly the fault of the employer then the liquidated damages clause would cease to bite and, as Emden's *Building Contracts* (8th edition, Vol. 1, p. 290) puts it

> 'the architect ... might have then no power to extend time so as to reimpose on the contractor the obligation relating to liquidated damages.'

Prior to this decision, the majority of specialist lawyers would have said that the clause 25.3.3 provision was *mandatory* and not directory. The Court of Appeal thinks otherwise.

Temloc Ltd v. *Errill Properties Ltd* is certainly an interesting case. It re-emphasises the need for care in considering liquidated damages. There is every reason why parties to building contracts should agree to liquidated damages for late completion, since proof of loss is often difficult.

If they agree on a figure, even a negative one, that will represent the totality of the employer's damages for late completion.

The employer who opts for liquidated damages does not have the option of general damages as an alternative. The parties are stuck with the agreed figure, even if it is '£NIL'.

A ceiling on damages?

The debate about liquidated damages was given a new slant by observations made in judgments in the unreported decision of the Court of Appeal in *E. Turner & Sons Ltd* v. *Mathind Ltd* (1986). This was a successful appeal against summary judgment for a contractor, where a contract incorporating JCT 63 terms appears to have been made orally and by correspondence, and the Bills made it clear that there was to be phased handovers of flat and garage units. The Bills and drawings gave the start and finish dates for each phase, but made no reference to liquidated damages. The Appendix entry to the unsigned JCT 63 Conditions showed liquidated damages as a rate of *£1000 per week*. The Sectional Completion Supplement had not been used.

On the contractor's alleged failure to meet the completion date for the phases – the architect having certified that the units ought to have been completed some months earlier – the employer deducted liquidated damages by dividing the £1000 a week by the number of phases and so arrived at a liquidated damages figure for each phase. This was not in accordance with the contract, nor was the employer's subsequent tactic of claiming £1000 for each phase. The liquidated damages provision was unenforceable because it was expressed in an inconsistent way: *Bramall & Ogden Ltd* v. *Sheffield City Council* (1983). The employer then argued that as expressed the liquidated damages were a penalty and that it was entitled to recover general (unliquidated) damages by way of counterclaim. This argument was based on *M J Gleeson Ltd* v. *Hillingdon London Borough* (1970), where the effectiveness of clause 12 of JCT 63 (now JCT 80, clause 2.2.1.) was upheld and so the Bill provisions had to be disregarded.

The Court of Appeal followed *Gleeson* but, in allowing the appeal, made some observations on the employer's entitlement to general damages for delay where liquidated damages prove to be unenforceable. These observations seem to fly in the face of other recent cases such as *Temloc Ltd* v. *Errill Properties Ltd* (1988).

Both Lord Justice Parker and Lord Justice Bingham were clearly of the view that a liquidated damages figure for the whole of the works did not preclude claims for general damages for breaches of contract such as the contractor's failure to meet phased completion dates, in contrast to *Temloc* v. *Errill* where the court ruled that liquidated damages were exhaustive of the employer's remedies for the breach of late completion. The contractor had argued that liquidated damages for the whole of the works excluded general damages for alleged failure to meet any phased completion obligations. Lord Justice Bingham did not agree:

'The plaintiffs may ultimately be held to be correct in advancing that construction; but it has this odd consequence. Even though *ex hypothesi* the earlier completion dates are binding ... the defendants would have no remedy in damages for breach of those terms. To achieve that result one would look for a clause excluding any right to damages for a breach which would, in the ordinary course, give a right to damages. The plaintiffs say there is such an exclusion in clause 22 (now JCT 80, clause 24). Again, that clause must be construed in the context of the whole contract ...; but it does not seem to me, standing alone, to be an effective exclusion of any right to damages for earlier breaches.'

Lord Justice Parker was even more enthusiastic about the employer's right to general damages, saying that:

'there is a perfectly good business reason for applying the liquidated damages only to the whole works, but no reason at all for saying that liquidated damages for any breach in that respect constitutes a ceiling to what may be recovered for failure to meet the successive handover dates'.

These remarks were, of course, not necessary for the decision in the case and so they are not binding authority for the proposition that the employer can have general damages for breaches of contract causing delay to phased handovers as well as liquidated damages for the whole of the works and that there is no ceiling equal to the amount of liquidated damages specified in the contract if the clause fails for technical or other reasons. If this is, indeed, the law, then it would be manifestly unjust. The very point of liquidated damages is that the amount specified is supposed to be a genuine and agreed pre-estimate of the employer's loss, recoverable without proof, and it is surely wrong that an employer should be able to have both liquidated and general damages in this sort of situation.

If the liquidated damages clause fails because of the employer causing delay not covered by the extension of time clause, a modern view certainly is that the employer cannot recover more than the rate of liquidated damages stipulated in the contract. As Emden's *Building Contracts*, 8th edition, Vol. 1, p. 289, puts it, this

'is on the principle that the employer having by the contract estimated the amount of injury delay would cause him, it would seem inequitable, when by the fault of the employer the provision as to liquidated damages has ceased to be applicable, that the contractor should be liable to pay a greater sum ... for delay'.

The point has not been decided as such. To allow the employer to recover more than the dead liquidated damages figure would be to give him an unjustified 'bonus' since the liquidated damages were a genuine pre-estimate of his likely loss judged at the date the contract was made, the figure being the agreed damages to cover the likely loss. What happens in the event is immaterial.

Now that this matter has been given an airing, it may be that we shall see more case law development and get a specific ruling on the point. The observations of their lordships in *Turner* v. *Mathind* are likely to take most of the construction industry by surprise and the approach of the differently-constituted Court of Appeal in *Temloc* v. *Errill* is to be preferred. Where a liquidated damages figure is inserted in a contract the reasonable assumption is that the agreed figure is intended to cover all damages for delay and the employer should not expect to recover a higher sum than that agreed when the contract was entered into if the clause fails for any reason. Any other view would enable the employer to profit from his own breach of contract and that flies in the face of the general principles.

Interest on liquidated damages

Once the architect has issued his certificate of delay under JCT 80, the employer is entitled to deduct liquidated damages at the prescribed rate from moneys due or to become due to the contractor. If, subsequently, the architect grants an extension of time so that a refund is due to the contractor, is the contractor entitled in law to make any claim against the employer for interest or financing charges which are attributable to the moneys refunded?

This was the point at issue in the Northern Irish decision of *Department of Environment for Northern Ireland* v. *Farrans (Construction) Ltd* (1982) and the judgment of Mr Justice Murray adds to the debate within the industry about the recovery of interest or financing charges in general.

The parties to the case agreed that, under a JCT 63 contract, the architect can issue more than one certificate of delay under clause 22 (JCT 80, clause 24), and so that important point was not at issue. Keating's *Building Contracts*, 4th edition, p. 345, suggests that more than one certificate can be issued:

'e.g., when the completion date is past and a certificate has been
issued but subsequently a cause of delay arises entitling the
contractor to an extension of time'.

Keating adds, in a footnote, that this is:

'because if the architect cannot grant a further extension the
original certificate would become bad and it follows that an architect
would never be safe in issuing a certificate until after completion'.

But this is an opinion which is not shared by all lawyers.

The real interest in *Farrans* case lies, however, in the fact that Mr
Justice Murray held that, in the circumstances set out, the contractor
was entitled to recover interest or financing charges, on the basis that
there had been a breach of contract by the employer, entitling the
contractor to damages, and the interest was recoverable as 'special
damage'.

The judge's reasoning is hard to follow, but appears to be
summarised in this passage of his judgment:

'It was the employer's own voluntary decision to make the
deductions and no doubt it took the decision in what it thought to be
its own financial interest. The only basis on which the employer can
justify the deductions ... is that there was a succession of dates on
which the Works ought reasonably to have been completed – each
date so to speak being valid and effective while the relevant
certificate stood – but I see nothing in the contract to justify such an
extraordinary conception and I reject it.'

With great respect to Mr Justice Murray there is nothing at all
extraordinary in that conception, which is, in fact, what the contract
plainly contemplates and, indeed, the argument appears to be that if
indeed several certificates of delay can be issued, the subsequent
certificates act retrospectively so as to make the employer's previous
deduction of liquidated damages (under the earlier certificate of delay)
a breach of contract retrospectively.

This criticism is supported by the commentary to the case by the
learned editors of *Building Law Reports* who rightly suggest that the case
must 'be regarded as of limited assistance outside Northern Ireland.'
Whether it would be followed by an English court is doubtful.

Having decided that, retrospectively, the employers were in breach
of contract by having deducted liquidated damages, Mr Justice Murray
went on to hold that the contractor could recover interest or financing

charges *as damages for the breach of contract*, as 'special damage,' this being an exception to – the common law rule against recovery of interest.

Mr Justice Murray relied on *Wadsworth* v. *Lydall* (1981) for the entitlement to special damages. There, the plaintiff agreed to sell a property to the defendant and, in anticipation of receiving £10 000 from his purchaser, entered into a contract to buy another property. The defendant (the purchaser) defaulted and paid a lesser sum very late. The Court of Appeal held that the plaintiff was entitled to recover interest and financing charges he had paid out as 'special damages'.

Mr Justice Murray said that, on the basis of the principle of *Wadsworth* v. *Lydall*, that if the contractor proves:

'(a) that because of interest or financing charges (he) suffered loss or damage through the employer's breach of contract in failing to pay any of the deducted moneys to the contractor on the due date, and (b) that a reasonable person in the employer's position would have contemplated such loss or damage as a probable consequence of the employer's breach of contract,'

then the contractor is entitled to damages to compensate him for that loss or damage.

No doubt contractors will continue to rely on the Northern Irish case to justify similar claims under JCT 80, but for many reasons these are probably doomed to failure. Even if the reasoning of Mr Justice Murray is accepted, the position under JCT 80 (and IFC 84) is different.

Clause 24.2.2 of JCT 80 provides that if the architect

'fixes a later completion date the employer shall pay or repay to the contractor any amounts recovered, allowed or paid under clause 24.2.1 for the period up to such later completion date.'

Nothing is said about interest or financing charges, and according to normal rules of contract interpretation, a gloss cannot be added to those plain words.

Hence, under JCT 80, no interest is recoverable in those circumstances – unless Parliament changes the law as to interest on overdue amounts in general. Furthermore, the working of clause 24.2.2 makes it equally plain that the employer is not making good an earlier breach of contract.

Clause 24.2.1 confers on him the right to deduct liquidated damages once the architect has issued his clause 24.1 certificate *and* the employer has given written notice of intended deduction to the contractor. Once these two conditions are satisfied, the employer has a contractual right to deduct liquidated damages – and clause 24.2.2 deals with what is to happen if the completion date is later altered in the contractor's favour.

The position is, in fact, the same under JCT 63. Few lawyers are convinced (as Mr Justice Murray was) that the employer is in retrospective breach of contract if a later certificate of delay is issued, and liquidated damages become repayable. If the employer, at the time when he deducts liquidated damages, is entitled to do so under the contract, that cannot possibly be a breach of contract under English law, and the *Farrans* case is not generally held to be sustainable in England.

Chapter 6

Sub-contracts

Direct contracts limit liability

Where a nominated sub-contractor enters into a direct but limited contractual relationship with the employer, the direct contract is inconsistent with any assumption of responsibility beyond that expressly undertaken.

So the Court of Appeal has ruled in *Greater Nottingham Co-operative Society* v. *Cementation Piling & Foundation Ltd* (1988), in allowing an appeal from the decision of the late Judge David Smout QC. Cementation was nominated sub-contractor for piling works to a main contractor operating under JCT 63 terms. It also entered into a direct employer-nominated sub-contractor agreement, in the then RIBA 'grey form' of warranty.

A restaurant adjoined the site in Skegness and, owing to the negligence of a Cementation employee, it was damaged. The employer met the restaurant owners' claims. Cementation did not dispute its liability to indemnify the employers, and piling works were suspended as a result. Consequent discussions on how to proceed led to delays, and eventually the architects issued instructions for a revised piling scheme.

The trial judge awarded the Co-op damages in both contract and tort. These included the additional cost of the revised piling scheme; the additional sums paid to the main contractor as direct loss and/or expense for the delay; and the Co-op's own consequential losses due to late completion. The appeal was concerned only with damages in tort since Counsel felt unable to support the judge's finding of liability in contract, although why this extraordinary concession was made is unclear, especially as the warranty contained an express undertaking by the sub-contractor that the main contractor would not be entitled to an extension of time 'by delay on the part of the sub-contractor'. The Court of Appeal thus expressed no view about the warranty, but the concession was obviously a vital factor in all three judgments.

Cementation's argument – which was successful – was that these heads of damage were not recoverable except for minor elements relating to physical damage, not relevant to the appeal. It was common ground that there was no breach of warranty under the direct contract between Cementation and the employer. The main contract extension of time and the additional cost of the varied piling work were dealt with in the ordinary way under the contractual machinery. The real point at issue was whether the so-called principle of *Junior Books Ltd* v. *Veitchi Co. Ltd* (1982), as discussed in subsequent authorities, should be extended or confined. There was the necessary close relationship between Cementation and the employers, but cases – such as *Muirhead* v. *Industrial Tank Specialities Ltd* (1986) – suggest that the structure of contractual relationships adopted may be inconsistent with co-terminus tort liability. Lord Scarman summarised the trend in *Tai Hing Cotton Mill Ltd* v. *Liu Chong Hing Bank* (1986):

> '[It is] correct in principle and necesssary for the avoidance of confusion in the law to adhere to the contractual analysis: on principle because it is a relationship in which the parties have, subject to a few exceptions, the right to determine their obligations to each other, and for the avoidance of confusion because different consequences do follow according to whether liability arises from contract or tort.'

What was the impact of the direct contract upon the otherwise very close relationship between Cementation and the Co-op?

Lord Justice Purchas observed that collateral contract:

> 'was restricted to the specific topics therein. If the parties intended a contractual restriction of the duty owed in tort, it would have been expected that specific provision would have been made in the contract ... The authorities established quite clearly that the duty in tort is capable of running coincidentally with the liability in contract. There is nothing in the terms of the collateral contract to destroy a liability which otherwise would have arisen in tort.'

This was the kernel of the employer's argument.

With some hesitation, Lord Justice Purchas found that Cementation's arguments were the more persuasive. He did not wish to open a 'Pandora's box'. Present policy was to restrict the *Junior Books* principle.

> 'Once it is established that there is no general liability in tort for pecuniary loss dissociated from physical damage ... it is difficult to

construct a special obligation of this nature in tort to which liabilities created by the collateral contract do not extend.'

The other members of the court agreed in allowing the appeal. Lord Justice Woolf discussed the economic damage issue in detail. He noted that the situation was very close to *Junior Books* but was unwilling to adopt the fiction that *Junior Books* was a case of physical damage. His lordship again regarded the direct contract as inconsistent with any assumption of responsibility beyond that expressly covered by the contract. He emphasised that this did not affect the nominated sub-contractor's normal liability in tort – it merely disproved the existence of the exceptional circumstances needed for liability for economic loss. The general principle now is that economic loss is not recoverable in the absence of physical damage to property or personal injury, except in a very special 'reliance' situation: *D & F Estates Ltd* v. *Church Commissioners* (1988).

Suppose, however, that a nominated sub-contractor does not enter into a direct collateral contract with the employer. He is simply nominated by the employer and contracts with the main contractor. Would a claim for mere pecuniary loss caused by his negligence lie? This example would seem to satisfy all the tests – the employer would be relying on the sub-contractor's skill and experience. There would be the necessary close proximity of relationship. This is, in fact, what appears to have happened in *Junior Books*. There was no actual privity of contract between the employer and the sub-contractor in that case, but it is to be treated as a case on its own special facts. In principle, therefore, the employer could not bring a claim direct in tort against the nominated sub-contractor to recover his foreseeble economic losses such as compensating the main contractor for delay and disruption, and himself for losses due to delayed completion.

Greater Nottingham Co-operative Society v. *Cementation Piling & Foundation Ltd* is very much a policy decision and it is unclear why the employer abandoned the claim in contract since it contained an express undertaking by the sub-contractor that the main contractor would not be entitled to an extension of time for 'delay on the part of the sub-contractor'.

Remarked Lord Justice Purchas:

'In this compartment of consideration it is not only the proximity of the relationship giving rise to reliance which is critical, but also the policy of the law as to whether damages for pecuniary loss ought to be recoverable.'

Policy has changed since *Junior Books*.

Policy for all

In any major building or engineering contract, where numerous different sub-contractors are involved, it is common practice for the main contractor to take out a single insurance policy covering all those involved in respect of loss of or damage to the entire contract works. This system has many advantages: it saves paperwork, avoids overlapping claims and – in the case of a very small sub-contractor – is more economic.

The legal effect of this practice was considered by the High Court in *Petrofina (UK) Ltd* v. *Magnaload Ltd* (1983), which arose out of an accident at the Killingholme oil refinery, where a major extension was being built. The main contractor was Foster Wheeler Ltd, but it had sub-contracted much of the work. The point at issue was whether two *sub*-sub-contractors were covered by an 'all risks' insurance policy taken out by Foster Wheeler, which defined 'the insured' as including the owner, the refinery operator, Foster Wheeler Ltd and/or contractors and/or *sub-contractors*.' Mr Justice Lloyd described the issue as 'a question of great importance for the construction industry generally.'

The insurer argued that the defendants were not sub-contractors as defined because at least one of them was a *sub*-sub-contractor. In contrast with the various indemnity cases – such as *Manchester Corpn* v. *Fram Gerrard Ltd* (1977) where a restrictive meaning has been given to the term 'sub-contractors,' the judge held that the defendants were 'sub-contractors within the ordinary meaning of the term'. So, the 'sub-contractors' in the context of the policy included *sub*-sub-contractors as well as ordinary sub-contractors.

The next question related to the property insured which the policy referred to as 'the works and temporary works ... [and] constructional plant ...' The judge refused to accede to the argument that each insured was only insured in respect of his own property. Business convenience, he said, meant that he must hold:

'that each of the named insured is insured in respect of the entire contract works, including property belonging to any other of the insured, or for which any other of the insured were responsible'

and this interpretation clearly equates with reality and commonsense.

Mr Justice Lloyd went on to deal with the nature of the all-risks policy, holding that it was a policy on property, pure and simple. He referred to the transport case of *Hepburn* v. *Tomlinson* (1966) where a firm of carriers took out insurance for goods in transit. A large quantity of cigarettes were stolen while still in transit because of the owner's negligence. The carrier was not liable to the owner but it made a claim under the policy for the benefit of the owner. Although underwriters had always regarded insurance in this form as being a liability policy as opposed to one on goods, the House of Lords rejected this view. The judge felt that the same principle was applicable to the case before him.

It covered 'all risks of loss or damage to the insured property' with certain defined exceptions. The final question for decision was a technical one, namely whether a main contractor can in law insure the whole contract works in his own name and the name of all his sub-contractors with the result that a sub-contractor could recover the whole of the loss insured, holding the excess over his own interest in trust for all the others.

Having looked at the case law – none of which involved sub-contractors – the judge felt that there was nothing in law to prevent this being done, notably because 'from a commercial point of view it has always been regarded as highly convenient'. The position of a sub-contractor in relation to contract works as a whole is sufficiently similar to that of a person to whom possession of goods is entrusted by an owner 'as bailee' in relation to those goods to enable him to hold, by analogy, that the sub-contractor is entitled to insure the entire contract works. 'In the event of loss,' he said, the sub-contractor is entitled 'to recover the full value of those works in his own name', subject of course to handing over any access to the others named.

Sub-contractors thus have an insurable interest in the whole of the contract works. Insurance provisions in the contract itself often give rise to serious difficulties of interpretation and, in practice, need to be determined by the contract provisions about transfer of risks.

Footnotes in the standard forms often point out the things the contractor should take into account. The majority of contract forms provide that the works are at the contractor's risk until practical completion. This means that the contractor needs to cover the possibility that the works are damaged by some cause which is not his fault at all such as theft or vandalism. In some cases, of course, insurance cover may not be available.

Although the wording of the corresponding clause has now been altered, the case of *A. E. Farr Ltd* v. *The Admiralty* (1953, provides a graphic example of the importance of scrutinising the contract wording. Clause 26(2) of the then current Government conditions of contract provided that the works were to be 'at the risk' of the contractor and made him responsible for making good 'any loss or damage thereto arising from any cause whatsoever ...' The contract works – a jetty at a Naval base – were near completion when seriously damaged by a destroyer colliding with the jetty. Despite the fact that the damage was caused by the employer's own negligence, in effect, Mr Justice Parker held that the contractor was obliged to repair and reinstate the work at his own expense. 'It all comes back to the meaning of the words in' clause 26(2), he said.

The *Petrofina* case was concerned with the wording of the policy itself; *Farr's* case concerned the meaning to be attributed to the wording of the construction contract. The problem in the latter case could have been avoided had insurance been taken out.

Compromised indemnity

Sub-letting work gives rise to many problems, whether or not the sub-contractor involved is nominated by the architect. The basic principle is that the main contractor is responsible for the acts and defaults of his sub-contractor so far as the employer is concerned. If the sub-contract work is defective, for example, it is to the main contractor that the employer will look for redress.

Standard contracts may attempt to alter the general position, thus JCT 80, clause 35.21, relieves the main contractor of liability to the employer for a nominated sub-contractor's design and allied failures, although the last sentence of the sub-clause emphasises that this limitation does not 'affect the obligations of the [main] contractor ... in regard to the supply of workmanship, materials and goods' by the sub-contractor.

If the main contractor is sued by the employer in respect of alleged defects in workmanship, materials or goods and claims that these are the responsibility of a sub-contractor – nominated or otherwise – he can bring the sub-contractor concerned into the legal proceedings as a third party, claiming indemnity from the sub-contractor in respect of the damages which he may be ordered to pay the employer and for his own costs.

Fletcher and Stewart Ltd v. *Peter Jay and Partners Ltd* (1976), is an interesting case for this very reason. It deals with the common situation where the main contractor compromises the claim brought against it by the employer and then seeks to recover the amount of the settlement (and legal costs) against the sub-contractor allegedly responsible.

The main contractor in this case made sub-contractors third parties to proceedings in respect of defective works, the action being settled by the main contractor paying £15 000 to the employer out of court. The main contractor did not obtain the sub-contractor's agreement to the settlement before making it, but subsequently pursued its third party proceedings against the sub-contractor.

The latter denied liability both for the defective work and for the amount of the settlement. All that the main contractor proved at the trial of the third party proceedings was the fact that the settlement was a reasonable one; it did not attempt to establish that the sub-contractor was substantially in breach of the sub-contract.

The Court of Appeal, upholding the trial judge, dismissed the main contractor's claim. It was for the main contractor to prove his case against the sub-contractor. The mere fact that the former had compromised the employer's proceedings on reasonable terms did not automatically entitle the contractor to recover against the sub-contractor.

This was so even though the sub-contractor had notice of the proceedings and the sub-contract terms imposed on the sub-contractor obligations identical to those borne by the main contractor to the employer under the main contract.

Lord Justice James described the main contractor's argument as a 'startling proposition'. The contractor ought to have proved that there had been a breach of contract by the sub-contractor resulting in the company having to pay over damages.

> 'The nature and amount of any settlement negotiated previously to that between the defendant and the plaintiff had nothing at all to do with the liability as between the defendant and third party.'

Apparently, the sub-contractor had admitted liability earlier as to a small sum.

This general position may, of course, be altered by express agreement between the parties, as was the position in *Comyn Ching and Co. Ltd* v. *Oriental Tube Co. Ltd* (1979), where there was what was described as a 'post-contract guarantee' between the parties.

The terms of this letter were held by the Court of Appeal to have created a legally-binding contract of indemnity which extended to liabilities or claims having a reasonable prospect of success. On the facts, it was reasonable to have compromised the employer's claim and the amount actually paid by the main contractor in settlement – amounting to some £9000 – was reasonable. The main contractor was held entitled to recover this amount, together with its own costs, under the terms of the indemnity.

The standard sub-contract forms, such as NSC4 and DOM/1, contain an indemnity clause, the wording of which is important in considering any claim which is being passed over to the sub-contractor. For example, had the contract in this case been in NSC4 form, Comyn Ching would not have had to resort to the post-contract guarantee except to the extent that the claim arose from its own negligence: see NSC4, clause 5.2.

The general legal position as regards compromised claims in a main contractor/sub-contractor situation is admirably summarised by the editors of *Building Law Reports* in the commentary on *Comyn Ching*: 'Where a claim has been settled it is normally necessary for a person seeking to recover the amount paid to establish in essence that there was legal liability to pay such an amount; that the amount paid was reasonable; and, of course, that the defendant is legally liable to pay such amount ... The amount of the settlement is *prima facie* evidence of the amount of loss caused by the defaulting sub-contractor.'

The fact remains, however, that if in third party proceedings the sub-contractor disputes liability, it is for the main contractor to prove the case against him. Once liability has been established (or admitted) then the amount of any compromise settlement is relevant evidence of the damages recoverable against the defaulting sub-contractor; but the amount of the compromise must have been reasonable.

Experience suggests that cases of this kind are not uncommon, but many people misunderstand the basic principles involved. The wording of the indemnity clause in a sub-contract may assist the main contractor but, in the standard forms commonly in use, the indemnity clause does not normally relieve the main contractor from its obligation to prove liability and damage. Indemnity clauses are, in any case, strictly interpreted: see *Walters* v. *Whessoe* (1977).

Certifying site delay

Nominated sub-contracts are a fertile source of disputes, and neither the general law nor the standard forms have yet resolved the tensions. A major practical problem is the way in which the main contract and sub-contract fit together, along with the role of the architect, when a main contractor alleges that a nominated sub-contractor is in delay.

The problem is probably less acute under JCT 80 than was the case under JCT 63, so far as the architect is concerned. JCT 63, clause 27(d)(ii) required the architect to certify a nominated sub-contractor's failure to complete the sub-contract works 'if the same ought reasonably so to have been completed' and also clause 27(d)(i) made the architect's written consent a condition precedent to the granting by the main contractor of an extension of time to the nominated sub-contractor. The difficulties for architects were highlighted by *City of Westminster* v. *Jarvis* (1970), as a result of which they had to apply the same test as the House of Lords in determining whether or not the sub-contract works were completed on time.

Many cases arising from JCT 63 are still coming forward, and in light of the *Jarvis* decision – which appears to be widely misunderstood (if not ignored) by the architectural profession – many clause 27(d)(ii) certificates are invalid because architects decided that a sub-contractor was in default without observing the ground rules laid down. It is quite immaterial that the sub-contractor could have worked faster – 'what matters is whether he has done what he agreed to do in the contractual time', said Lord Wilberforce.

JCT 80 has introduced complex provisions relating to nominated sub-contractors but, on examination, clause 35 is likely to produce as many problems as it purports to solve. Clause 35.15.1 is the equivalent of JCT 63, clause 27(d)(ii), and requires the architect, if notified by the contractor that a nominated sub-contractor has failed to complete on time, to 'so certify in writing to the contractor'. The key-phrase is 'If any nominated sub-contractor *fails* to complete ... within the period specified in the Sub-Contract' as extended by the main contractor with the architect's consent. The architect must be satisfied that there has been a failure to so complete by the sub-contractor and, before issuing his certificate of default, he must be satisfied that any relevant extensions of time have been granted. In fact, the clause says 'provided that ... clause 35.14 has *been properly applied'* that clause requires him 'to operate the relevant provisions' of the sub-contract as regards extensions of time.

In NSC4, extensions of time are covered by clause 11.2, which largely parallels clause 25 of JCT 80, although a nominated sub-contractor is also entitled to extensions of time for delays caused to him by defaults of the main contractor and those for whom he is legally responsible. In such a case the contractor is to fix, with the architect's consent, such new sub-contract period as the architect considers reasonable.

It is stated that *in agreement with the architect* the contractor is to state in his extension (a) which matters they have taken into account and (b) the extent, if any, to which the architect has had regard to omission instructions issued by him. Nothing is said as to what is to happen if contractor and architect do not agree – which is not a remote possibility bearing in mind the fact that the contractor is being asked to adjudicate upon his own default. There will be some nice problems for arbitrators in the future!

Reverting to the architect's certificate of default under clause 35.15, without that certificate the main contractor has no right to claim against the sub-contractor for failure to complete the works on time.

124 Some Building Contract Problems

The clause 35.15 certificate is simply a statement of fact that the sub-contractor has not completed within the specified time. This is in contrast with JCT 63, clause 27(d)(ii) where the certificate was a statement of opinion – that the sub-contractor 'ought reasonably to have so completed'. But the nature of the architect's duty in issuing the certificate is much the same under both editions of the contract. JCT 80 places on the architect a specific obligation to satisfy himself that the sub-contract extension of time provisions have been properly operated and in fact gives him up to two months to make the necessary inquiries: see clause 35.15.2. This runs from the date of the contractor's notification of alleged failure to complete.

Before certifying, the architect needs to make quite exhaustive inquiries – though probably the burden upon him is not so great as was the case under JCT 63. If the sub-contract documentation has been correctly completed – itself a gargantuan task in view of its complexity and the possibility for error – the architect ought not to be faced with the sort of problems commonly thrown up by 'Green Form' nominations. So, for example, part II of the 'Green Form' appendix entry relating to completion period was often completed vaguely; e.g. 'in accordance with main contractor's overall programme' or, even worse, 'within a period to be agreed'. This sort of thing meant, in effect, that the architect could not be satisfied of the precise date of completion which meant that he could not – or should not – certify that the sub-contractor was in default.

That sort of problem is avoidable under the current scheme of nomination – provided all the parties complete the documentation in the proper way. Nonetheless, architects, main contractors, and nominated sub-contractors alike would do well to read and digest the speeches in *City of Westminster* v. *Jarvis* if the acrimonious and costly disputes of JCT 63 days are to be avoided and, in particular, study the observations of Lord Wilberforce. Contractors and architects must give careful consideration – before sub-contracting – to such questions as the sequence of the sub-contract operations and their effect on the main contract works as a whole, and similar practical problems which the standard documentation grasps only imperfectly. Tender NSC1 is often not fully and properly completed – the vital schedule 2 is often annotated 'to be agreed' – which makes the scheme inoperable.

Named complications

The JCT Intermediate Form of Contract 1984 (IFC 84) contains, in clause 3.3, provisions for 'named persons as sub-contractors'. They are complicated and unclear and throw up many problems. Indeed, clause 3.3 gives the impression that its terms have not been thought through.

A named person can come about by being mentioned in the contract documents; the work is to be priced by the contractor but carried out by the person named. Alternatively, a person can be named in an architect's instruction about expending a provisional sum. In this second case the contractor has a right of 'reasonable objection'.

In the first case, clause 3.3.1 says 'the contractor shall, not later than 21 days after entering into this contract' sub-contract with the named person using Section III of the Form of Tender and Agreement NAM/T.

Quite apart from the fact that three weeks is a wholly unrealistic period, the date of 'entering into this contract' is not necessarily the date it was signed. It could also be 21 days from the date on which the employer accepted the contractor's tender – which is a very different thing.

The draftsman clearly foresaw difficulties. 'If the contractor is unable so to enter into a sub-contract *in accordance with the particulars given in the Contract Documents*' he is to inform the architect immediately. He must specify 'which of the particulars have prevented the execution of such sub-contract'. Exactly what 'particulars' are referred to is not clear – but presumably the intention is to refer to such things as programming, etc. Even the notorious JCT 80 nomination documentation is clearer than this!

Architects are faced with problems, too. It is for the architect to complete Part I of Form NAM/T and it must be implied that this duty will be carried out by the date of 'entering into this Contract'. If – for good reason – the architect is unable to meet that requirement, the contract is silent on what is to happen.

Potentially the contractor might have a claim at common law – he cannot do what clause 3.3.1 requires unless that documentation is ready.

There is no power to award extension of time on this ground – such delay is not, for example, a late instruction – and the tardy architect's only remedy would be to advise the employer to defer giving possession of the site under clause 2.2 (if it applies) with the inevitable consequence that the contractor gets both extra time and extra money.

Assuming both contractor and architect know what the 'particulars' are, and they prove a stumbling-block to the sub-contract, the architect must issue instructions if he is satisfied that those 'particulars' have prevented the making of the sub-contract. The architect's instructions can:

- change the particulars so as to remove the impediment; *or*
- omit the work; *or*
- omit the work from the contract documents and substitute a provisional sum.

The first two constitute a variation, with the financial consequences flowing from that. The third case requires the architect to issue a further instruction under clause 3.3.2, under which, among other things, he can require the work to be carried out by another named person when the provisional sum instruction must describe the work to be done and include the relevant details from NAM/T *and* the contractor has a right to object to the person named – assuming that he has reasonable grounds for so doing.

It is then for the architect to sort out the contractor's objection – the reasonableness of which can be tested in arbitration – and the only realistic approach at this stage would be based on negotiation. Under the provisional sum approach, the administrative burden is great – and the contract does not say what is to happen if the contractor does make a reasonable objection to the person named.

Effectively, it seems that the architect must either omit the work by way of a variation instruction or – by analogy with the duty to re-nominate under JCT 63 and JCT 80 terms – name someone else. The ensuing delay must be paid for – I assume that the architect can instruct postponement of the work under clause 3.15, thus giving rise to an extension of time (clause 2.4.5) and, where appropriate, to loss and/or expense under clause 4.12.5.

But let us assume – as we must – that all goes well, and a sub-contract is duly entered into.

IFC 84 makes complicated provisions for what is to happen if things go wrong, and the consequences depend on whether the named person's work was included in the contract documents or the naming came into being through an instruction about a provisional sum.

If the named sub-contractor fails – and that word is used broadly – the position appears to be that the architect must issue instructions following the failure which may either name someone else to execute the outstanding work, or tell the contractor to make his own arrangements for it, or omit the work entirely.

Where the work was originally included in the contract documents, and the architect takes the first option, the contractor gets time but not loss and/or expense, although the contract sum is adjusted to take account of any increased price quoted by the new named person. Under the second and third options the instruction gives rise to a claim for both extension of time (clause 2.3) and direct loss and/or expense (clause 4.11), and also ranks as a variation.

In those cases where the named person was created through a provisional sum instruction, the contractor gets both time and money, and the instruction also ranks as a variation. In both cases this is subject to an overall proviso – if the contractor caused the failure of the named subcontractor, e.g. by wrongfully terminating the sub-contract, etc. he cannot benefit from his own default.

These comments only touch the surface of part of this long and complicated clause, which seems to be an unhappy compromise between the various conflicting interests.

Direct payment problems

Most standard form construction contracts contain direct payment clauses under which the employer is entitled to pay nominated sub-contractors directly if the main contractor fails to pay them. JCT 80, clause 35.13 is a typical provision. Direct payment is discretionary unless the employer has entered into a direct warranty agreement with the nominated sub-contractor. It is then mandatory (clause 35.13.5.1).

The subject is both important and controversial. Some lawyers take the view that direct payment clauses fall foul of the general principles of insolvency law. They rely on the House of Lords' decision in *British Eagle International Airlines Ltd* v. *Compagnie Nationale Air France* (1975), which, so the argument runs, overruled by implication the well-known building industry case of *Re Tout and Finch* (1954), which clearly established that the employer is entitled to pay sub-contractors direct if the main contractor becomes insolvent. The better view is that *Re Tout and Finch* is still good law because of the fiduciary nature of retention funds.

The sub-contract in that case contained a clause which expressly declared that sub-contract retentions were to be held by the main contractor as 'a trustee for the sub-contractor.' This ruling has been upheld in many recent cases on building contracts.

The first decision is that of Mr Justice Vinelott in *Rayack Construction Co. Ltd* v. *Lampeter Meat Co. Ltd* (1979), where the contract was in JCT 63 form. The judge granted an injunction against the employer requiring it to pay the retention monies into a separate bank account. The basis was that clause 30(4) of the contract was designed to protect the contractor in case of the employer's insolvency. Unusually, there was a 50% retention which was to be held for the whole of a five-year defects liability period!

The judge had no doubt about the point. In his view, the retention clause read in light of the contract as a whole imposed:

> 'an obligation on the employer to appropriate and set aside a separate trust fund ... equal to that part of the sum certified as due in any interim certificates in respect of work completed which the employer is entitled to retain during the defects liability period.'

The protection was extended further in *Re Arthur Sanders Ltd* (1981), where a trust of the retention monies was implied in favour of a sub-contractor even though the employer had failed to set aside a separate trust fund in accordance with the *Rayack* principle. The sub-contractors were in the old NFBTE/FASS 'green form'. Most importantly, Mr Justice Nourse expressed a very clear view that the *British Eagle* decision (that direct payment could not take place) did not affect the position. He said:

> 'In order that there may be no doubt about it ... The decision of the House of Lords in *British Eagle* ... has nothing to do with this kind of case.'

The more recent decision in *Re Jartray Developments Ltd* (1983) does not affect the position at all. That case emphasises that it is not sufficient simply to make provision for the creation of a trust fund. A trust must actually be created. Although under the contract the employer was bound to set aside the retention as a separate trust fund, it had not done so. And the contractor had not applied to the court for an injunction requiring it to do so. It had missed the boat.

Mr Justice Nourse held that this remedy ceased to be available to the contractor once the employer went into liquidation. As no part of the money had been appropriated or set aside before liquidation started, the contractor could not claim to be treated as if it had been, and it does not appear to have been argued that 'Equity treats what should have been done as having been done' – for which there is copious authority apparently not cited to the court.

The case is important because it shows what can happen if a contractor or sub-contractor does not take appropriate steps to protect his interests at the right time.

In *Jartray* there was a development contract and a JCT contract. In the earlier cases there were JCT main contracts and NFBTE/FASS sub-contracts. 'In the latter kind of case,' said the judge, 'there is a very close linkage between the building contract and the sub-contract'. As he pointed out, clause 11(h) of the old 'green form' took effect as an express assignment to the sub-contractor of a due proportion of the contractor's beneficial interest in the retention monies under the main contract. This was not so in the *Jartay* case. This is a point which sub-contractors working under non-standard or in-house forms should check.

In fact, clause 27 of JCT 80 radically altered the wording of JCT 63, clause 25.

Commenting on this change the *JCT Guide* p. 24, says:

'In clause 27.4.2, second sentence, the employer's right to pay any supplier or sub-contractor is now excluded if the reason for the contractor's employment being determined [terminated] is bankruptcy or liquidation; the reason for this change ... is that the 1963 provision was open to challenge by the ... liquidator of the contractor and it was considered ... inappropriate to include it in the 1980 edition'.

A distinguished London solicitor – a specialist in building law – has stated firmly that this amendment stems 'from an erroneous view of the law' and with his view I am in entire agreement. The amendment was based on the wrong assumption that the *British Eagle* case 'had somehow overruled Mr Justice Wynne-Parry's decision in *Re Tout & Finch*'. This is not so, the recent cases have all proceeded on the contrary assumption that direct payment can take place, but the fact that there are differing views indicates a measure of uncertainty. Construction industry insolvencies remain high, and so the point is a vital one. The building industry line of cases has consistently maintained what was always the law before *British Eagle*. This must be a comfort for most main and sub-contractors.

Doing the 'subbie's' work

One of the main criticisms of JCT 80 – levelled by architects and contractors alike – relates to the extremely complex provisions on nominated sub-contractors, contained in clause 35. There is certainly force in those criticisms, insofar as the 'basic method' of nomination is administratively complex and time-consuming, but so long as the industry wishes to retain the system of nomination, these problems will remain.

It has, however, been suggested that clause 35 of JCT 80 has made a significant inroad into the main principle established by the House of Lords in *North-West Metropolitan Regional Hospital Board* v. *T. A. Bickerton and Son Ltd* (1970).

The House of Lords laid down that if a nominated sub-contractor defaults or fails – whether by reason of insolvency or otherwise – the architect must make a fresh nomination.

More recently, the House of Lords has said that the architect's duty to re-nominate must be exercised 'within a reasonable time' of the main contractor seeking a re-nomination instruction: *Percy Bilton Ltd* v. *Greater London Council* (1982).

A nominated sub-contractor's failure is not a fault or breach of contract for which the employer is responsible in law, and so the main contractor bears the immediate financial consequences of the sub-contractor's failure. This is, perhaps, a good reason for main contractors to insist on a performance bond.

The *Bickerton* principle is now enshrined in JCT 80. Clause 35.24 imposes an express contractual duty on the architect to re-nominate, subject to the observance of the somewhat convoluted procedural provisions. That clause distinguishes between two kinds of failure by a nominated sub-contractor: (1) sub-contractor's default in specified respects; (2) financial failure.

In the first case, the reference is to 'default in respect of any one or more of the matters referred to in clause 29.1.1 to .1.4 of the sub-contract NSC4 or NSC4a as applicable' – the use of that form of sub-contract being mandatory.

It has been suggested that, as a result, this has eroded the second principle in the *Bickerton* case, for the House of Lords made clear that the main contractor has no right or duty to carry out nominated sub-contract work himself.

As Lord Reid said in *Bickerton*, to hold otherwise would

'be contrary to the whole purpose of the scheme, and it would be strange if the [main] contractor could have to do work for which he never tendered and at a price which he never agreed.'

The argument that this fundamental rule has been breached in JCT 80 runs as follows. NSC4, clause 4.5, deals with the sub-contractor's failure to comply with the main contractor's directions. It says that if after written notice to comply, the sub-contractor does not so comply, then with the permission of the architect, the main contractor may *employ and pay other persons* to comply with such direction' at the defaulting sub-contractor's expense. It is said that in some way this provision gives the main contractor the right to carry out the nominated sub-contract work himself: see (1982) *Chartered Quantity Surveyor*, p. 299.

This is placing a strained and unreasonable interpretation upon the italicised phrase which is not, according to normal principles of interpretation, to be read as including the contractor and his own employees, at least so far as doing the *whole* of the nominated sub-contract works goes.

Obviously, it would be within the clause for the main contractor to instruct his own operatives to carry out some specific task, in respect of which a main contractor's direction had been given. If there is a substantial default by the nominated sub-contractor, then the determination provisions of NSC4 or 4a, clause 29 would apply.

There seems to be no conflict between NSC4, clause 4.5, and any provision of the main contract; and certainly the principle laid down by Lord Reid in *Bickerton* is still good law and is not to be put to death by a side-wind.

A second limb of the argument is that, under JCT 80, clause 25.4.7, whereby the main contractor may be entitled to an extension of time for 'delay on the part of nominated sub-contractors ... which he has taken all practicable steps to avoid or reduce', coupled with the contractor's obligation to use 'his best endeavours to prevent delay in ... progress ... howsoever caused', the main contractor may have to do the work himself if he is to meet these requirements.

The better view is that these words require no more than a general obligation 'to show willing'. Certainly, the 'practicable steps' which the contractor must take do not, it is clear, extend to executing the nominated sub-contract work himself, for that is something that he has no power or right to do.

This is, indeed, made plain in JCT 80 (and implied in JCT 63) by clause 19.5.2:

'Subject to clause 35.2 the contractor is not himself required to supply and fix materials or goods or to execute work which is to be carried out by a nominated sub-contractor.'

If an argument along the above lines is advanced when asking for permission under NSC4, clause 4.5 for the contractor to carry out the work himself, the prudent architect should reject it. The argument has a certain superficial attractiveness, but does not appear to be sustainable either on general principle or as a matter of interpretation of the contract as a whole.

And as Mr Justice Forbes remarked, speaking of a different JCT clause, 'one has to construe the contract as a whole': *Algrey Contractors Ltd* v. *Tenth Moat Housing Society Ltd* (1972).

The general scheme and intent of the provisions relating to nominated subcontractors is plain. What Lord Reid said in *Bickerton* in 1970 about JCT 63 is true today of the JCT 80 provisions:

> 'The scheme creates a chain of responsibility ... but I can find nothing anywhere to indicate that the principal contractor can ever have in any event either the right or the duty to do *any* of the [work reserved for nominated sub-contractors].'

Entitlement to claims

Sub-contractors are at the sharp end of the construction industry and no subject is more hotly disputed than a sub-contractor's entitlement to claims, despite the fact that the standard form sub-contracts are very clear in their terms. Problems arise when the sub-contract is in a non-standard form (usually of the main contractor's own devising), but under the industry-agreed forms the position is clear.

Under NSC4, for example, the nominated subcontractor has two basic entitlements to money claims:

- Clause 13.1 gives him the right to claim against the employer *through* the main contractor for direct loss and/or expense in respect of seven 'matters' which are, in law, the employer's responsibility. The main contractor must obtain these benefits for the sub-contractor, provided the correct procedure is followed.
- In contrast, clause 13.2. entitles the sub-contractor to the value of any direct loss and/or expense which results from the default or omission of the main contractor or those for whom he is responsible in law. The acts or omissions must disturb the regular progress of the sub-contract works to a material extent.

Similar, though less detailed, provisions exist in related sub-contract forms, and it is only in the area of 'in-house' main contractors' forms that real problems are likely to arise. Even here, in some cases a sub-contractor may be able to bring a claim against the main contractor for disturbance to the regular progress of the sub-contract works on the basis of breach of an *implied* term that the main contractor will not hinder or prevent him in the carrying out of the sub-contract works although it follows from the most recent cases that there will seldom be room for terms to be implied where there is a detailed written contract, unless without the implied term the contract will not work.

Whether such a term can be relied upon will depend on the actual terms of the sub-contract, and in many cases the express terms will leave no room for a term to be written in: *Martin Grant & Co. Ltd* v. *Sir Lindsay Parkinson & Co. Ltd* (1985). In that case, the sub-contractor had undertaken to carry out his work 'at such time or times as the [head] contractor shall direct'. This wording left no room at all for any term to be implied and meant that the sub-contractor had no claim even though he had to carry out the work several years later than he had originally envisaged. The sub-contractor had undertaken the risk that the main contract and the sub-contract might go on much longer than was originally contemplated, and was simply unwise in not protecting himself against that risk.

This sort of problem will not arise under any of the standard-form negotiated sub-contracts, since each one dovetails with its main contract equivalent. The main difficulty under the standard forms is that of proving the various heads of claim. The onus of proof rests on the sub-contractor who must establish, by adequate evidence, that he has suffered direct loss and/or expense as a result of one or more of the specified matters, as well as the amount of his claim.

Figures cannot be plucked out of the air and the sub-contractor must be able to produce back-up by way of site records, invoices and the like. If he is claiming 'loss of productivity', for example, in a prolongation situation he will need adequate documentary evidence and, in my view, will need to demonstrate that his original tender was realistic and show the difference between his tender figure and the labour hours actually expended.

Greater difficulties may arise where the claim is against the main contractor on the basis that the loss or expense is caused by his 'act, omission or default'. In that case, the standard forms envisage that the amount of the claim will be *agreed* upon between main and sub-contractor and this is often easier said than done.

Under the sub-contracts the sub-contractor's duty is merely to give written notice to the main contractor within a reasonable time of the occurrence of the disruptive event. Agreement of the amount of the claim is presumably by negotiation. No main contractor is going to agree any *claim* unless hard evidence is produced.

Although the sub-contracts are silent on the matter, in practice a fairly detailed claim document is required, with supporting documentation. There is no ideal format, but if the claim is not settled the dispute could well end up in arbitration. It is therefore sensible to produce the claim document with that in mind, setting out the basis of the claim, exact details of the events relied upon (related to programme if necessary) and detailed calculations.

Sub-contractors' claims against main contractors cover a wide legal spectrum. 'Act, omission or default' of the main contractor brings in common law claims as well and common law principles apply to their settlement.

If the claim is not settled, the sub-contractor's remedy is to go to arbitration and in the case of a small claim that will seldom be worthwhile.

Claims by main contractors *against* sub-contractors are fraught with equal difficulty and entail as many pitfalls. However, two common fallacies can be knocked on the head. In the first place, where the sub-contractor is nominated under JCT terms, the main contractor has no claim against the sub-contractor for late completion unless the architect has issued a certificate of delay: *Brightside Kilpatrick Engineering Services Ltd* v. *Mitchell Construction Ltd* (1975). This is a point sometimes overlooked by main contractors.

Similarly, under the standard forms, the main contractor's right of set-off is limited to what the sub-contract terms provide and in general does not extend to setting off claims arising under one contract against payments due on another. This is the situation under the JCT-related forms but is often altered by an amending clause or in home-devised conditions. Main and sub-contractors alike must study the sub-contract wording and then operate its procedures correctly.

A set-off dispute

A main contractor's right of set-off against a nominated sub-contractor working under NSC4 was the subject of a decision of the Court of Appeal in *Archital Luxfer Ltd* v. *A. J. Dunning & Son (Weyhill) Ltd* (1987). The court ruled that summary judgment may be given on part of a claim and the balance of the action stayed pending arbitration, not only if an arguable defence is raised, but also if mixed questions of law and fact are at issue.

Dunning was main contractor for the erection of a hostel at Basingstoke and Archital was nominated as sub-contractor for the supply and fixing of aluminium windows. The sub-contract was in the JCT standard form NSC4, 1980 edition, clause 23.2, which sets out the main contractor's right to set-off where the amounts are not agreed. It provides that:

(1) No set-off relating to any delay in completion shall be made unless the architect has issued his certificate of delay.
(2) The amount of such set-off has been quantified in detail and with reasonable accuracy by the contractor.
(3) The contractor has given to the sub-contractor notice in writing specifying his intention to set off the quantified amount.

These are pre-conditions to the right to set-off. Clause 24 then provides that if the nominated sub-contractor disagrees with the amount which it is intended to set off, he can refer the matter to the adjudicator, with his reasons for disagreeing.

Archital claimed in respect of sub-contract monies certified by the architect to be due to it. These were not paid because Dunning claimed that it was entitled to set off a greater amount for expense incurred as a result of delay by Archital. It was common ground that, subject to this possible set-off, Archital was entitled to £26 516. The company applied for summary judgment under Order 14 of the Rules of the Supreme Court.

Before His Honour Judge John Newey, QC, Official Referee, they contended that Dunning had failed to comply with the pre-conditions laid down by clause 23.2 and that Dunning had no 'arguable defence'.

Judge Newey held that Dunning's answer that it had complied with clause 23.2 was at least arguable, except for the first of the arhitect's certificates, which was for £1541. He gave Archital judgment for that sum. He held that Dunning was entitled to leave to defend for the balance, and directed that the balance be stayed pending arbitration between the parties.

The Court of Appeal dismissed Archital's appeal and ruled that the judge was plainly right. Lord Justice May referred with approval to views expressed by Lord Justice Kerr in the *Sethia Liners* case (1985). He said that if defendants raised an arguable point of law which the court was able to consider without referring to contested facts, it would consider it to see whether there was any substance in the proposed defence. In the majority of cases where there is an arbitration clause, the whole dispute should then be referred to arbitration. The presence or absence of an arbitration clause was not generally relevant when a court was considering whether there was any substance in the alleged point of law: see *Chatbrown Ltd* v. *McAlpine Ltd* (1986).

Lord Justice May agreed with this view, but Archital's case raised mixed questions of law and fact. Dunning relied on a letter from the architect as constituting the necessary delay certificate. It referred only to 'delay to the North Block' and was in very general terms. There was also some suggestion of a verbal agreement relating to the completion date and the usual sort of arguments in this type of case.

There was also dispute about whether the amount had been quantified. Dunning argued that it had, by a letter making a provisional claim for liquidated damages and direct loss and/or expense. This was a provisional assessment. 'As a matter of law,' ruled the Court of Appeal, 'liquidated damages can never be passed on by a main contractor to his sub-contractor.' The letter did, however, contain at least an arguably sufficient quantification of Dunning's loss within the requirements of clause 23.2.2.

The main purpose of the clause 23 procedure is to enable the sub-contractor to operate the adjudication provisions of clause 24, under which he may send a written statement of disagreement to the contractor and the adjudicator and so get the dispute resolved. Looked at in that light, Lord Justice May thought that it was at least arguable that Dunning's letter was sufficient to enable Archital to answer the claim and get the problem resolved by adjudication.

Mixed questions of law and fact are, as the court indicated, preeminently matters for the expertise of an arbitrator rather than that of the court. It could not be said that Dunning's defence to a substantial part of the claim was unarguable.

This is a sensible and realistic decision, though I suspect that it will not be welcomed by sub-contractors! But it accords with the scheme of the sub-contract, and is in line with all the recent authority.

The existence of the arbitration clause does not prevent the sub-contractor from going for summary judgment when he sues. As Lord Denning MR said in *Associated Bulk Carriers Ltd* v. *Koch Shipping Ltd* (1978), 'the court ought to give judgment for such sum as appears to the court to be indisputably due and refer the balance to arbitration.' From the sub-contractor's point of view, of course, this is unsatisfactory. But it is what the sub-contract envisages. Main contractors are faced with similar problems when suing on main contract architects' certificates.

The great to-do about set-off will no doubt continue, even though the so-called principle in *Dawnays Ltd* v. *F. G. Minter Ltd* (1971) is dead and buried, although there are still those who contend otherwise. Unless – and it is not likely – the whole payment scheme is radically changed, payment problems and set-off arguments will continue to bedevil both the industry and the courts, as is borne out by the increasing number of cases going before the courts, e.g., *Smallman Construction Ltd* v. *Redpath Dorman Long Ltd* (1989), where a purported set-off notice was again found to be defective.

Chapter 7

Disruption and Prolongation Claims

The secret of successful claims

Many contractors' claims for loss and/or expense fail because they are not backed up by supporting evidence. Figures cannot be plucked out of the air, and it is the contractor's job to provide the necessary evidence to prove his case. All too often, claims are made and settled on the basis of an *assessment* whereas, under the JCT contracts at least, the duty of the architect or quantity surveyor is to *ascertain* the amount of the loss and expense which the contractor has suffered or incurred.

The JCT forms place a heavy burden on the architect or quantity surveyor in a claims situation, and JCT 80 requires the contractor to provide details of his claim. The contract wording is important, and contractors have only themselves to blame if they fail to respond to the architect's request for details. None of the standard form contracts require the contractor to submit a 'claim' in any particular form, although the ICE Conditions of Contract for Civil Engineering are much more precise than other contract forms.

Claims normally fall into two groups: prolongation and disruption, but the principles involved are the same. The contractor's entitlement is to recover the actual loss and/or expense which is directly caused by the event giving rise to the claim. It is not sufficient for the contractor to establish that the delay or disruption would not have occurred had it not been for the event. The loss or expense must be caused directly by the event, e.g. late information. Lawyers call this the principle of causation. A contractor's failure to show that the loss or expense is directly caused by the specified event is a ground on which a claim may be rejected.

No two claims are alike. In each case the contractor can recover his actual demonstrable losses and/or expenses, and in many cases these are easy to establish. In a prolongation situation, for example, site costs are relatively easy to establish during the overrun period, whereas 'head office overheads' or 'establishment' charges are not and so, rightly or wrongly, the contractor resorts to the formulae put forward in the textbooks. The use of formulae can be avoided if there is sufficient data to establish the contractor's exact loss and/or expense.

If a formula has to be adopted, many quantity surveyors favour that set out in Emden's *Building Contracts and Practice*, 8th edition, Supplement, Volume 2, p. n/46:

$$\frac{h}{100} \times \frac{c}{cp} \times pd$$

In this formula h is the head office percentage which is arrived at by dividing the total overhead costs and profit of the contractor's organisation as a whole by his turnover in the relevant period. This can be established by reference to the contractor's audited accounts. The contract sum is represented by c, while cp is the contract period and pd is the period of delay.

This is more realistic than the Hudson formula in its approach and it may be used with back-up evidence such as the audited accounts. Certainly, head office overheads are a permissible head of claim if backed up by acceptable evidence. The best evidence is the cost of resources employed at head office during the period of prolongation as demonstrated by the contractors' records.

'Loss of productivity' is also easy to prove in principle, although it is difficult to establish in detail, and all too often such a claim (if admitted at all) ends up as an assessment. Indeed, some writers have argued that there is no entitlement to a claim under this head, but for reasons which I have set out more fully in *Building Contract Claims*, 2nd edition, pp. 139–140 with John Sims, I believe that it is allowable, even if it is difficult to establish in practice. Once again, some quantity surveyors will accept a formula method of evaluation.

One item which is never a permissible head of claim is the cost of preparing it. Certainly, under the JCT contracts it is the duty of the architect or quantity surveyor to ascertain the amount of the claim, based on the detailed information submitted by the contractor.

The situation is different if a claim proceeds to arbitration, because in that event the arbitrator might properly award an outside specialist's fees as costs of an expert witness: *James Longley & Co. Ltd* v. *South West Thames Regional Health Authority* (1983). In that case, a claims consultant had been involved in an arbitration.

On the other hand, if the architect or quantity surveyor requires the contractor to do more than the contract requires – for example by carrying out special research which would not normally be done – in principle this would be allowable as a head of 'special damage' in accordance with the rule laid down by *Tate & Lyle Food and Distribution Co.*

Ltd v. *Greater London Council* (1982), where a High Court judge held that the cost of managerial time spent in remedying an actionable wrong could properly form the subject-matter of a claim for 'special damage' in a common law action.

From the contractors' point of view, the secret of successful claims is to keep proper records, and to do what the contract requires. Happily, where JCT 80 is used, there is a very clear procedural code, and under that form one popular head of claim – 'finance charges' – has become of less importance than it was under JCT 63. The 1980 contract makes it abundantly clear that claims are to be dealt with as the contract proceeds and not 'spiked' until job completion. Nonetheless, the case law makes plain that the contractor is entitled to recover, as an integral part of his direct loss and/or expense, the financial burden to him of financing the loss or expense, or the loss of possible interest which he would have earned had the money been available to him to invest. But the contractor must prove his loss here as in other cases.

Balancing the probabilities

I am sometimes taken to task for stating that in a claims situation the contractor must prove his claim on the balance of probabilities. It is said that this is against the whole principle of reimbursement of loss and expense which is to ensure that the contractor is actually reimbursed for whatever costs he has sustained as a direct result of the relevant matter.

The position is not, in fact, so simple. Under JCT terms the phrase used is 'direct loss and/or expense' which is not the same as 'costs'. It may be that there is some scope for interpretation of those words 'direct loss and/or expense' and in what may be said to arise as a direct result of 'regular progress of the works' being materially affected but in effect it gives two dual heads of claim:

• Actual losses incurred as a direct result of the event relied on.
• Actual expenditure or disbursements similarly occasioned.

If it is accepted that the money claims clauses in building contracts are regulated provisions for the payment of damages, to harp a well-worn string, then common law principles must apply to the ascertainment.

One of those principles is that it is for the claimant to establish or prove his claim on the balance of probabilities. If he has accurate and detailed cost records, all well and good, but how many contractors are in that happy position?

Happily, the whole area of JCT money claims was clarified by *London Borough of Merton* v. *Stanley Hugh Leach Ltd* (1986). That case confirms that if the contractor makes a written application for the reimbursement of direct loss and/or expense and the architect forms an opinion favourable to him, the ball is then in the contractor's court. Mr Justice Vinelott said:

> 'The contractor must clearly co-operate with the architect or the quantity surveyor in giving such particulars of the loss or expenses claimed as the architect or quantity surveyor may require to enable him to ascertain the extent of that loss or expense'.

He added emphatically that if the contractor:

> 'fails to answer a reasonable request for further information he may lose any right to recover loss or expense under those sub-clauses ...'

In the *Merton* case, the contractor's calculation of direct loss and/or expense came under both judicial and arbitral scrutiny. It totalled £2 245 605, and commenced with 'direct site costs', from which clause 31A fluctuations were deducted, a percentage for overheads and profit was then added back, together with the net fluctuations so as to arrive at the total figure.

This calculation made no allowance to the employer for the financial risks which the contractor was bound to shoulder under the contract, e.g. delays which do not merit extensions of time, but was no doubt based on the contractor's cost records.

The arbitrator, as quoted by the judge, pointed out that very few building contracts proceed without a hitch and only in that ideal situation would the calculation provide a direct comparison with the tender figure.

Mr Justice Vinelott found it 'impossible to see how the calculation ... can be treated as even an approximation of a claim, whether or not rolled up ... under clauses 11(4) and/or 24(1) of JCT 63'. This was on the assumption that the contractor was able to prove the breaches alleged.

The reference to a 'rolled up' claim is, of course, to *Crosby* v. *Portland Urban District Council* (1967) which Mr Justice Vinelott followed. He added that it was implicit in the decision

> 'that a rolled up award can only be made in a case where the loss or

expense attributable to each head of claim cannot in reality be separated and secondly that a rolled up award can only be made where apart from that practical impossibility the conditions which have to be satisfied before an award can be made must have been satisfied in relation to each head of claim.'

The purpose of the contract claims machinery is to put the contractor back into the position he would have been but for the delay or disruption, and various cases establish that the term 'direct loss and/ or expense' means in effect that what is recoverable is substantially the same as the damages recoverable at common law under the first rule in the old case of *Hadley* v. *Baxendale* (1854).

The reference to 'proof on the balance of probabilities' is to the standard required in English law and that is what architect or quantity surveyor are entitled to insist upon. The contractor's duty is to provide the architect or quantity surveyor with information supporting his claim and if he fails to do so then, as the *Merton* case emphasises, he may lose out. In ascertaining a claim, architect or quantity surveyor must treat it in the same way as a judge or arbitrator would in awarding damages at common law. Cost records provide the best starting point but are not the final yardstick because applying common law principles may demonstrate that only part (or none) of the costs are in fact reimbursable.

All too often contractors' claims are presented under the contract on the basis of the difference between the contractor's actual expenditure and his actual recovery, as was apparently done in the *Merton* case. In some cases, of course, this may be the correct approach as no two claims situations are identical, but in the majority of cases it comes perilously close to a *quantum meruit*.

The architect's duty is a relatively simple one when dealing with money claims. Once he has decided that the contractor's application is valid, he (or the quantity surveyor if that duty is assigned to him) must find out for certain the amount of the claim. This he can only do on the basis of information provided by the contractor. It has been described as a relatively mechanical exercise, but it is one which cannot be performed unless the contractor is able to produce satisfactory evidence to prove each and every head of claim.

Money claims under JCT 80

Money claims under JCT 80, clause 26, (and its equivalent in other JCT forms) are a controversial topic and employers sometimes allege that the philosophy of some contractors is 'tender low, claim high'. In fact, claims under clause 26 are closely circumscribed. The contractor must comply with the procedural mechanism of the clause and he must establish that he has suffered or incurred '*direct* loss and/or expense' as a result of the event relied on. Clause 26 confers a *right* to extra payment provided the contractor relies on it and its mechanism is observed.

One of the areas of controversy results from the fact that loss and expense claims under JCT forms have traditionally been linked to extensions of time. This erroneous view (see *H. Fairweather & Co. Ltd* v. *London Borough of Wandsworth* (1988)) is compounded by clause 26.3. This has the side note 'relevance of certain extensions of completion date'.

In truth, there is no connection between clause 26 and the extension of time provision, clause 25. An extension of time under clause 25 does not entitle the contractor to any extra payment, whether by way of 'extension of preliminaries' or otherwise. Claims under clause 26 are quite separate and distinct. The lack of connection between extensions of time and claims for direct loss and/or expense was emphasised in *H. Fairweather & Co. Ltd* v. *London Borough of Wandsworth* (1988) where it was expressly held that the grant of an extension of time under what is now clause 25 is not a condition precedent to a claim for direct loss and/or expense under clause 26.

Clause 26 speaks of 'regular progress of the Works or any part there-of' being 'materially affected' by one or more of the events listed in clause 26.2. Some legal authorities argue that this must necessarily involve *delay* in progress, thus cutting out loss of productivity as a head of claim.

This view is put forward by I. N. Duncan Wallace in his *Building and Civil Engineering Standard Forms*, p. 112, where he says that:

'it is limited to loss or expense due to *delay to progress* and does not cover cases where the work, without progress being affected, has become more expensive or difficult.'

He adds that this is unfair and that the phrase, no doubt, 'may be interpreted liberally in the contractor's interest, but it is clear that it does not cover loss of productivity'.

With respect, this line of argument is not convincing. 'Progress' means 'move forward or onward; be carried on', and it seems plain that 'regular progress of the works' can be 'materially affected' without there being any delay to completion at all – the word 'delay' is not used in the clause.

Wallace himself provides a good example of the sort of thing that, in his view, does not give rise to a claim and contrasts it with the situation under JCT 63, clause 11(6), under which he allows that loss of productivity is a head of claim.

A late delivery of a vital instruction may well cause the contractor to incur additional cost because he cannot use the labour and plant resources available to him in an efficient way. Late instructions amount to a breach of contract in any event and a claim under clause 26.2.1 is not limited to heads of damage caused by delay to progress. It covers other possible consequences of the late instruction, including loss of productivity.

Claims under clause 26 can be equated with claims for damages at common law, and are governed by the same principles. On that basis, the consequences of the late instruction must have been within the reasonable contemplation of the contracting parties at the time the contract was made. The additional cost arises as a direct and natural result of the architect's default and is thus claimable under the contract terms: see *Hadley* v. *Baxendale* (1854).

The contractor must, of course, be able to *prove* loss of productivity, and this may be difficult to do. But assuming that he can provide the necessary evidence, it is thought that loss of productivity is a valid head of claim.

Clause 26 permits claims for both *disruption* and *delay* (or prolongation). A claim under the first head arises where the employer (through his architect or otherwise) makes work more difficult or expensive than foreseen; a prolongation claim arises where the work is actually delayed and the completion date is affected.

In practice, it is often difficult to separate the two types of claim because claims for both often arise from the same facts. Indeed, the position may be further complicated by the presence of extra work, i.e. variation work.

How is the architect or (more usually) the quantity surveyor to ascertain the direct loss and/or expense in such circumstances? It is up to the contractor to provide him with the necessary information from records. However, there are very considerable difficulties in calculating the precise loss arising from both delay and disruption under this and, indeed, others heads.

Emden's *Building Contracts and Practice*, 8th edition, Vol. 2, p. N/45 puts forward a method of calculation.

'Initially, a period is examined when the contract was running normally, and the value of work done during that period is assessed and then divided by the number of operatives and/or items of plant on site. The figure thus arrived at is compared with the same figure calculated for the period of delay or disruption, and the comparative figures are then used to calculate the amount of loss.'

This, of course, has the merit of simplicity, its legal basis is doubtful. Quantity surveyors asked to approve the use of such a method should, obtain the specific authority of the employer to settle on this basis.

As a matter of law it is for the contractor to prove his claim on the balance of probabilities and the duty of the architect or quantity surveyor is to 'ascertain' the amount of direct loss and/or expense. 'Ascertain' is defined in the dictionary as 'find out (for certain), get to know' and is different in meaning from 'assess' or 'estimate'.

Some people argue that the actual financial effect of loss of productivity is difficult, if not impossible, to 'ascertain'. They then argue that in the absence of precise 'ascertainment' a reasonable *assessment* should be made based upon the balance of probabilities. The argument has some attraction. In practice the use of such calculations and (in other areas) formulae of one kind or other may be used, subject to limitations. But in essence claims settled under the contract are equivalent to damages at common law and so must be assessed in accordance with common law rules.

Claims cannot be 'spiked'

The Court of Appeal upheld the landmark decision of His Honour Judge John Newey QC in *Croudace Ltd* v. *London Borough of Lambeth* (1985), and in so doing knocked many common fallacies on the head. It is now clear that an architect who 'spikes' contractor's claims potentially renders his employer liable in damages for breach of contract, and the fact that the architect has not certified any amount as being due in respect of the claim will not prevent recovery of it. The courts will protect a contractor's rights even in the absence of a certificate.

Furthermore, in the Court of Appeal, Lord Justice Balcombe expressed doubt about an employer's right to insist that all claims must be approved by him before payment. His lordship noted that outside quantity surveyors, who were acting as agents for the council's chief quantity surveyor,

'who had certain independent duties to perform under the contract ... Unless [a letter] is intended to mean that on this matter Lambeth's chief quantity surveyor was not prepared to delegate his functions under the contract to the coutside firm, it is not clear to me what right Lambeth had to issue this instruction.'

In fact, with great respect to Lord Justice Balcombe, no employer has that right; the architect or quantity surveyor is charged with the performance of independent functions under the contract and any interference with those functions is a legal wrong for which the contractor has a remedy in damages.

It appears from the Court of Appeal judgments – though not apparent from the first instance report – that the architect and quantity surveyor had been instructed not to pay money claims 'until amounts have been agreed and cleared by the [council's] audit section, which is allegedly the usual practice of both local and health authorities. In law the audit section has nothing at all to do with it; it is the architect who is charged with the duty of ascertaining the amount of any direct loss or expense incurred by the contractor – though he may delegate that duty to the quantity surveyor – and once amounts are so ascertained they are to be certified and paid. That is what the contract clearly says, whether public bodies like it or not.

Croudace had sought summary judgment for an interim payment of £100 000 plus interest, this being the sum recommended as an interim payment by the quantity surveyors. Alternatively, it claimed the same sum as damages. The basis of Croudace's claim was that Lambeth was liable to pay because it was responsible as employer under the contract or, alternatively, as employer of the architect who had failed to ascertain Croudace's claim for disturbance to regular progress of the works.

The full amount of Croudace's claim under the contract was for £646 994.63, but the Court of Appeal ruled that the contractual claim could not be maintained in the absence of a certificate from the architect. But that did not prevent Croudace from recovering damages for Lambeth's breach of contract in failing to instruct the ascertainment of Croudace's loss and expense. This they had done by failing to renominate an architect under the contract when the original named architect retired.

His Honour Judge Newey QC had ruled that the JCT scheme of things was for the contractor's loss and expense to be ascertained by the architect and included in the next interim certificate. The failure of the architect to perform his duties – or if he was prevented from performing them by the employer – amounted to breaches of contract for which the employer was liable.

In the Court of Appeal it was accepted that Judge Newey's further holding that there was implied in a JCT 63 contract an obligation on the part of Lambeth to nominate a successor architect when the named architect retired. Consequently that was a further breach of contract by Lambeth. As Lord Justice Balcombe noted, Lambeth's acts and omissions including but not limited to its failure to nominate a successor architect amounted to a failure by it to take such steps as were necessary to enable Croudace's claim for loss and expense to be ascertained, and as such amounted to a breach of contract on its part: *Smith* v. *Howden Union* (1890).

The real issue before the Court of Appeal was whether Croudace could establish that it had suffered damage resulting from the breach of contract and, if that issue was answered in the affirmative, whether the damage should be quantified by the court or in arbitration.

On the first point the court was quite clear: 'Unless it can be successfully maintained by Lambeth that there are no matters in respect of which Croudace is entitled to claim for loss and expense ...' it necessarily follows that Croudace must have suffered some damage as a result of there being no one to ascertain the amount of its claim. Lambeth could not discharge that onus and so Croudace was home and dry on that point.

The Court of Appeal was firmly of the opinion that the trial judge had been right to refuse Lambeth's application that the legal proceedings be stayed and allowed to go to arbitration. The court felt that Lambeth had no defence on liability, but merely on amount, and this was a factor that should be taken into account. A payment on account of damages was quite appropriate, and there was the general question of Lambeth's conduct. Its conduct in failing to take steps necessary for the ascertainment of Croudace's claim merited 'the strongest condemnation'.

Lord Justice Balcombe said:

'Moreover the court is entitled to infer, and I do so infer, that the object of this conduct was to postpone the evil day when Lambeth would have to pay to Croudace the amount which the architect had acknowledged was due, even if its precise amount had not been quantified. Further, it is a reasonable inference that a motive, if not the sole motive, of the application to stay proceedings was to create further delay.'

This unmeritorious conduct was a very relevant factor and the court concluded that Judge Newey was right to refuse a stay and right to order immediate payment of £100 000 on account.

Privilege not to be abused

Public policy encourages disputing parties to settle their differences rather than litigate them to a finish. This is the reason for the 'without prejudice' rule which excludes all negotiations genuinely aimed at reaching a settlement from being given in evidence in any subsequent arbitration or litigation. In 1988 the House of Lords ruled that the privilege is not lost if a settlement is reached, with the effect that disputing third parties are not entitled to see the settlement correspondence.

This was the key point in *Rush & Tompkins Ltd* v. *Greater London Council* (1988), which arose out of a dispute on a building contract. Rush & Tompkins were the GLC's main contractor. Careys were ground works sub-contractors who submitted claims for loss and expense to Rush & Tompkins. The GLC did not agree Careys' claims. Rush & Tompkins started proceedings against both the GLC and Careys in respect of Careys' entitlement and claiming reimbursement from the GLC. There was a settlement between Rush & Tompkins and the GLC, and while its terms were disclosed to Careys they did not show what valuation had been put on the claim.

Careys believed that the negotiating correspondence must have shown the value of their claims for purposes of the settlement. They applied for discovery of the without prejudice correspondence i.e. its disclosure, but the Official Referee refused it accepting Rush & Tompkins' argument that the documents were protected by the without prejudice rule because they came into existence for the purpose of settling the claim. The Court of Appeal reversed his decision and ordered discovery holding that the privilege ceased once a settlement had been reached.

The House of Lords overturned this ruling and restated the law. Lord Griffiths emphasised that the public policy which protects genuine negotiations from being admissible in evidence must be extended to protect those negotiations from being disclosed to third parties. This is of great importance in a contract claims situation where the employer and main contractor reach a compromise which is unacceptable to a sub-contractor. Lord Griffiths put the issue succinctly:

'The wiser course is to protect without prejudice communications between parties to litigation from production to other parties in the same litigation. In multi-party litigation it is a not infrequent experience that one party takes up an unreasonably intransigent attitude that makes it extremely difficult to settle with him. In such circumstances it would place a serious fetter on negotiations between other parties if they knew that everything would be revealed to the one obdurate litigant. What would in fact happen would be that nothing would be put on paper but this is in itself a recipe for disaster in difficult negotiations which are far better spelt out with precision in writing'.

Claims negotiators can take heart and rely on the 'without prejudice' privilege. It is best to head all negotiating correspondence 'without prejudice' to make clear that in the event of the negotiations being unsuccessful they cannot be referred to in any subsequent proceedings. In fact, the privilege does not depend on the use of the phrase. If it is clear from the surrounding circumstances that the parties were seeking to compromise a dispute, the negotiations cannot be used subsequently to establish an admission or partial admission of liability.

However, the privilege is not absolute. In certain limited circumstances without prejudice material can be looked at if the justice of the case so requires. The House of Lords noted some of these exceptional cases:

- Without prejudice material can be looked at if the issue is whether or not the negotiations resulted in an agreed settlement: *Tomlin* v. *Standard Telephones & Cables Ltd* (1969).
- The court will not allow the phrase to be used to exclude an act of bankruptcy nor to suppress a threat if an offer is not accepted: *Ex parte Holt* (1892); *Kitcat* v. *Sharp* (1882).
- Without prejudice correspondence can be looked at to determine the question of costs after judgment or award: *Cutts* v. *Head* (1904).
- The admission of an independent fact in no way connected with the merits of the dispute is admissible even if the admission was made in the course of negotiations for a settlement, e.g. an admission that a document is in the handwriting of one of the parties.

These are, however, all exceptional cases. They cannot be used to whittle down the protection given to negotiating parties. The vital point is, though, that 'without prejudice' statements and discussions will only be privileged if there is a dispute and an attempt to settle or compromise it. One should not head letters 'without prejudice' indiscriminately on the mistaken assumption that this gives one an opportunity to write anything with impunity. It does not.

But where there is a dispute, e.g. as to a contractor's entitlement to extension of time and/or money , and negotiations are in progress, the privilege should be invoked. Any admissions made cannot then be used in evidence in any subsequent proceedings connected with the same subject matter unless both parties consent. Admissions made to reach settlement with a different party in the same proceedings are also inadmissible whether or not a settlement is reached with that party.

The House of Lords' decision was welcomed by the legal profession and by all those engaged in disputes settlement. The fact is that whatever the standard form contracts may provide about payment of claims for disruption or delay, the typical scenario on a long contract is for the global claim to be settled by negotiation at the end of the day, especially where some of the delay is the contractor's responsibility. Both common sense and the law go hand in hand in protecting the negotiations even if they prove abortive.

Varied work may affect recovery of expenses

Contractors working on Government contracts under GC/Works/1, Edition 2, terms know that the claims clauses refer to 'expense' and not to 'loss and/or expense' which is the JCT terminology. Clauses 9(2)(a)(i) and 53(1) refer to the contractor 'properly and directly (incurring) any expense' beyond that otherwise provided for in the contract or reasonably contemplated by it.

This would limit claims to those cases where money is actually spent so that loss of profit, for example, would be excluded.

There is no case law directly in point, but the better view is that claims under GC/Works/1 are not limited to a 'money paid out' basis. The Property Services Agency (PSA) (which is by far the major user of the form) seem to accept this interpretation since in its Notice to Tenders (C2041, July 1986), the PSA agrees that 'expense' includes 'interest not earned if the contractor uses his own capital', and the PSA has formally accepted since 1984 that financing charges are recoverable, in principle, under Edition 2 of GC/Works/1.

However, in practice, the PSA's implementation of that decision has the effect of limiting a contractor's effective right of recovery of finance charges. This does not accord with the general principles of law enunciated by the Court of Appeal in *F. G. Minter Ltd* v. *WHTSO* (1980), and clarified in *Rees & Kirby Ltd* v. *Swansea City Council* (1986).

The problem arises because of the PSA's interpretation of clause 9(1) of GC/Works/1. The PSA accepts that where any prolongation or disruption is caused by a matter referred to in clause 53(1), the contractor is entitled to recover 'expense', including finance charges, under that clause. But at the same time the PSA interprets clause 53(1) (when it relates to variation instructions) as being limited to the disruptive effect of variations on other non-varied work. The PSA approach is that the value of the varied work itself (including disruption and financing charges) is to be valued under clause 9(1). Quite correctly, PSA says that because there is necessarily a delay between the execution of any work and the contracting being paid for it, the contractor's original rates must be taken to include an element for finance charges to cover that delay.

How this principle is applied is a matter of disagreement between contractors and the agency. While contractors must accept that, because there will inevitably be some delay between the execution of all work and payment, the contractor must be deemed to have allowed in his tender for financing the period of delay, the length of that period is open to question.

The contractors' view is that it must be the case that such a delay may not extend beyond a 'reasonable' period and that, accordingly, where the valuation of a variation is not undertaken within such a 'reasonable' period, the contractor cannot be deemed to have included finance charges to finance the extra cost of the varied work for the period of unreasonable delay before the certificate is issued. In short, the contractor cannot possibly be deemed to have included in his rates for financing the varied work over an indeterminate, longer period which he cannot have anticipated since it was caused by a variation order. In that regard, it is my firm view that financing charges over that longer period are properly recoverable under condition 53(1).

The contractors' argument seems convincing in light of the case law.

Contractors are also at variance with the PSA in relation to financing charges on under-valuation and an extended retention fund. This can be illustrated by two examples:

- A contractor carried out a variation which the quantity surveyor promptly values under clauses 9(1)(b) at rates deduced from the bill rates. The contractor wants a 'fair' valuation under clause 9(1)(c), and a year later the quantity surveyor agrees and increases the valuation, including an element of financing charges equivalent to those included in the bills. That element falls short of the actual financing charges incurred. It seems that the shortfall is recoverable under clause 53(1).
- The quantity surveyor wrongly fails to include the value of billed work in an interim valuation and/or the Superintending Officer wrongly fails to include the value in an interim certificate. The value is included in a much later certificate.

In that case, it has been suggested that the contractor is still entitled to recover the cost of financing the delay in payment, not under clauses 9(2) or 53(1) but on the basis that there was a breach of contract by the Authority. GC/Works/1 does not exclude common law claims either expressly or by implication. Two alternative arguments are advanced to support this view. First this was not interest on late payment of a debt because no debt comes into being until the certificate is issued. The common law rule against interest on late payment of a debt does not apply to breach of an obligation other than one simply to pay money: *TransTrust SPLR* v. *Danubian Trading Co. Ltd* (1952).

The second and more convincing argument is that the quantity surveyor's failure to carry out valuations in accordance with the contract or the failure of the Superintending Officer to certify in accordance with the contract amounts to a breach of contract for which damages are recoverable: *Perini Corporation* v. *Australian Commonwealth* (1969), and the first instance decision of Judge John Newey QC in *Lubenham Fidelities* v. *South Pembrokeshire District Council* (1983). Whether any contractor will be brave enough to challenge the PSA's refusal to meet such a claim is one matter; but once again the argument is an impressive one, with regards both to under-valuation and to delayed release of the reserve where delay is caused by a clause 53(1) matter.

Lost contribution must be proved

One of the most controversial elements of any contractor's claim for 'direct loss and or/expense' is the use of formulae for ascertaining the head office overhead (and profit) element in a contractor's prolongation claim under JCT and GC/Works/1 contracts. Certainly, the *uncritical* use of any particular formula is never advisable and the limitations of the formula approach must be noted.

These important heads of claim are concerned with the consequences of head office resources being tied up where a contract is prlonged thus being unable to generate a head office and overhead contribution on another contract elsewhere.

This was the situation in the well-known Canadian case of *Ellis-Don Ltd* v. *The Parking Authority of Toronto* (1978), where the plaintiff had been delayed for some 32 weeks on a car park contract. It was found that 17½ weeks of the delay was due to the employer's failure to obtain an excavation permit, and had been thrown into winter working as a result.

The High Court of Ontario ruled that the contractors were entitled to recover as damages for breach of contract: the extra cost of pouring concrete in winter; and the onsite costs for the 17½ weeks.

A weekly sum for overheads and loss of profit calculated by

reference to 3.87% which had been built into the tender which was assessed, in essence, by reference to the 'Hudson formula'.

The court accepted that there was no logical distinction to be drawn between a claim for lost profit and a claim for lost head office overheads. However, there are several crucial points in the judgment of Mr Justice O'Leary which are often conveniently overlooked by those seeking to rely on the Hudson formula without at least some evidence to back it up, even after the mathematical inaccuracy in it is corrected. [As normally used, the formula gives double-recovery, since the contract sum already includes the HO profit percentage.]

The judge said:

'Where a lump sum amount, in this case 3.87% of the total contract price, is included for profit and overhead, I am aware of no reason why the plaintiff should not recover the full amount of it, *where the evidence indicates the staff tied up by the ... delay would otherwise be earning such an amount for the plaintiff*.'

The emphasised words are important, because unless it can be proved on the balance of probability that contribution would have been earned elsewhere, a claim under this head must fail.

Another key passage in the judgment is also frequently ignored. It reads:

'That is not to say that the plaintiff would have been entitled to claim the lost profit on a contract it could otherwise have had if such a contract was lost by the ... delay. It is one thing to say that the parties when the contract was entered into should have contemplated that there was a real danger or serious possibility that staff tied up by the defendant's fault beyond what would have been the date for completion would be unable to earn normal profit and overhead for the plaintiff elsewhere. *It is quite another thing to say that the parties should have contemplated that the profit from a particular contract would be lost because of delay in completion caused by the defendant*.'

Here, the judge is referring to the principle that the contractor is only entitled to recover such part of the resultant loss as is reasonably foreseeable as likely to result from the breach.

This is judged at the time the contract was made, and so normally precludes the contractor from recovering profit allegedly lost on another contract.

The crucial finding in *Ellis-Don* was that

'there was ample evidence tendered before me that the men tied up for an extra 17½ weeks through the fault of the defendant could have been readily employed elsewhere and could have continued to earn overhead and profit for the plaintiff in the amount of Can $2,445.40'.

(The total award for the head office/profit percentage claim was Can $42 794.50.)

In simple terms, therefore, the Hudson formula was backed up by hard evidence. The actual method by which such loss is calculated appears to be very much less important than establishing the position of the deprivation of overheads and profit due to continued retention of staff.

An alternative method of checking a contractor's claim for head office overheads and profit is discussed by Geoffrey Trickey in the book *Presentation and Settlement of Contractors' Claims* (1983), pp. 123–132. He also summarises the criticisms of the Hudson formula as normally relied on. This method is illustrated by worked examples. On the same figures, the Hudson approach produces a total of £6918.64, while the 'adjusted' approach separates head office overheads and profit and produces £5450.21.

Any formula is bound to be a broad-brush method, and the ideal situation, seldom found in practice, is to have detailed cost and accounting records directly related to the particular contract. The main point is that the contractor's claim should not overstate the actual loss. This applies whether Hudson or its variants are used.

The commentary on *Ellis-Don* in *Building Law Reports* (Vol. 28, p. 103) puts the matter squarely:

'Whether the claim is presented as damages for breach of contract or upon a contractual right to recover loss, expense or damage the person liable for the claim, if established, or those acting on his behalf will wish to see all the documents relevant to checking the claim such as the make-up of the tender, and the internal head office and job records and accounts relating to overheads and to actual and anticipated profits and profit levels both in and on either side of the delay. In litigation or abitration such documents will be disclosable ...'

In summary, a formula is only as good as the evidence backing it up. The use of a backed-up formula simplifies the task of both parties, but it is not a panacea for all ills. Intelligently used, the Hudson formula, and its competitors, no doubt have a place.

Summer into winter working

Old law cases and the principles which they establish can often be used with good effect in the changed conditions of today and so it is with the case of *Bush* v. *Whitehaven Port and Town Trustees* (1888), which can be relevant when claims arise under the standard forms of contract.

One aspect of the case has been disapproved, though not overruled, by the House of Lords in a later case, but it has also been followed by a very strong Court of Appeal.

The facts have a familiar ring. Bush contracted with the Trustees to lay a 15 inch water main from Ennerdale Water to Whitehaven for £1335. The contract was made in June, and the Trustees contracted to be ready at all times to give Bush sufficient possession of the site to enable him to proceed with the works.

In fact, through the fault of the Trustees, the whole of the site was not available until October. As a result of this delay, the contract was thrown into the winter months, and Bush was put to heavy extra expense for which he sued the defendants.

His claim was successful, despite the fact that there was an express term saying that 'non-delivery of the site ... shall not ... entitle the contractor to any increased allowance in respect of money ...'

In one sense the decision is unsatisfactory, since the Court advanced alternative bases for its decision, namely that a summer contract having, by implication, been in the contemplation of the parties when the contract was made, Bush was entitled to *quantum meruit* ('as much as it is worth') or damages in respect of the increased expenditure.

As explained in the later case of *Sir Lindsay Parkinson Ltd* v. *Commissioners of Works* (1949), the decision is based on an implied term about the circumstances in which the contract works were to be done, and this certainly seems to have been the view of Lord Coleridge CJ in the *Bush* case itself. He dealt with the question whether a term could be implied for breach of which the plaintiff could recover damages, and was inclined to answer it in the affirmative, and referred again and again to the contract being made 'with reference to certain anticipated circumstances' and becoming inapplicable to the actual situation.

Here, what was to have been a summer contract had been turned into a winter contract, and Lord Coleridge is worth quoting at length.

'It was turned into a winter contract – into a contract when wages were different ...; when days were short, instead of long; when weather was bad, instead of good; when rivers which had to be dealt with, and had to be crossed by the pipes, were full not empty; and when, in fact, ... a great many most important circumstances under which the contract was to be executed, had ... changed from those which ... were in the contemplation of the parties when the contract was entered into. The contract, nevertheless, was carried on and completed ... with the knowledge of the defendants ... the contractor [completed the works] under the altered conditions ...'

This principle seems to be applicable to many cases arising under JCT 80, ICE 5th edition, and other standard form building contracts. Part or late possession of the site is given – which is itself a breach of the express or implied terms of the contract, unless the employer has the power to defer the giving of possession – but the contractor soldiers on.

In a refurbishment contract, it is assumed that houses will be given to the contractor in sequential groups but, because the employer cannot re-house the tenants, this is not done.

In these circumstances, it is likely that extra cost will be incurred, and it must be recoverable under the *Bush* principle, or under an implied term that possession will be given in groups: see *Whittal Builders Ltd* v. *Chester-le-Street District Council* (1988), where a term for giving of possession of houses in pairs was implied.

Even more commonly, variations are ordered, or there are late architect's instructions, which have the effect of turning a summer contract into a winter one. Any 'direct loss and/or expense' incurred as a result is, subject to proof, recoverable: and *Bush* v. *Whitehaven Trustees* can safely be used in support.

Obviously, the prudent contractor will allow a contingency or float in his programme against delays which are his responsibility, but he is not bound to take the totally unforeseeable into account. Prolongation of a contract which means working through an additional winter period almost inevitably results in extra 'cost' or 'loss and/or expense' to the contractor, e.g. extra costs of extending protective measures to materials, or replacing them due to deterioration, labour disruption, 'extended preliminaries' and so on, subject to any compensating saving caused by carrying out late work in better weather conditions than anticipated.

Such claims are often rejected by architects, engineers and quantity surveyors; they ought not to be, subject to the contractor providing the necessary information and, of course, satisfying the criteria of the claims provision itself.

If not recoverable under the express terms of the contract, e.g. JCT 80 clause 26, because (for example) the contractor has failed to make his written application at the right time, the loss may still be recoverable at common law, by way of litigation or arbitration.

Paraphrased extracts from the case of *Parkinson* v. *Commissioners for Works* are often – and wrongly – appended to contractor's claims for reimbursement of loss and expense in a prolongation situation. That case has no relevance to such claims. It was a very special case, and a reading of the full report illustrates the immense difficulties involved – there was a deed of variation to the original contract and the payment based on cost plus profit under the deed was much less than would have been due under a valuation in accordance with the terms of the original contract.

Bush's case was considered in *Parkinson*, discussed *and* applied, but it was a case involving substantial *additional works* done at the employer's request.

The 'summer into winter contract' situation is different and so requires a different approach.

Heads of claim are easily written but may be difficult to substantiate: the basic principle is that all the costs, etc claimed must be directly attributable to the cause relied upon by the contractor.

Late information

One of the most regular complaints from contractors is that the architect is late in supplying information. Indeed, so common is this problem that the JCT contracts make specific provision for it, and provide a contractual remedy, by way of reimbursement of direct loss and/or expense (JCT 80, clause 26.2.1) and an appropriate extension of time (JCT 80, clause 25.4.6).

At a recent conference – somewhat astonishingly – the point was raised that clauses 25 and 26 were limited to instructions for which the contractor must apply under clause 2.3.

That clause, of course, deals with discrepancies in, or divergencies between, certain of the contract documents and is conditional:

'*If the* contractor shall find any discrepancy in or divergence between any two or more of the [stated] documents ... he shall immediately give to the architect a written notice specifying the discrepancy or divergence, and the architect shall issue instructions in regard thereto'.

There is certainly no positive duty imposed on the contract to go through the documents looking for discrepancies, and equally there is nothing in clauses 25.4.6 or 26.2.1 that limits their application to a clause 2.3 situation.

Contractually, the situation is very simple. Clauses 5.3.1 and 5.4 require the architect to supply to the contractor such further information as he needs to supplement the contract drawings in order to carry out and complete the works. Even if the obligation were not express, the architect would be under an implied duty to supply necessary further information at a reasonable time: *Neodox Ltd* v. *Borough of Swinton & Pendlebury* (1958).

There, dealing with a situation which had arisen under a civil engineering contract, Mr Justice Diplock was quite specific:

'To give business efficacy to the contract, details and instructions necessary for the execution of the works must be given by the engineer from time to time in the course of the contract if he fails to give such instructions within a reasonable time, the corporation are liable in damages for breach of contract'.

Under JCT terms, the architect's failure to supply the necessary information required at a time which will allow the contractor to make use of it in order to complete the works on time is a breach of contract for which the employer is vicariously liable and for which the contractor is entitled to damages. But the architect is not bound to supply information in accordance with a contractor's programme which shows a completion date earlier than the contact completion date: *Glenlion Construction Ltd* v. *The Guinness Trust* (1987). JCT 80 confers on the contractor the right to have the breach of late supply of information dealt with by means of an extension of time and/or recovery of direct loss and/or expense *provided* that he has made a *specific* written application to the architect for it. A general application is not sufficient.

This point was one of the matters in issue in *London Borough of Merton* v. *Stanley Hugh Leach Ltd* (1985), where the High Court concluded that the submission of a programme early in the contract could constitute the necessary application, provided it met the contractual requirements about timing.

Mr Justice Vinelott pointed out that the purpose of the provision was in part

'to ensure that the architect is not troubled with applications too far in advance of the time when they will be actually needed by the contractor (thus disrupting unnecessarily the work of the architect and his staff) and to ensure that he was not left with insufficient time to prepare them'.

He emphasised that if in fact the work does not progress in accordance with his programme, 'some modification may be required to the prescribed timetable and the subsequent furnishing of instructions and the like'.

The structure and intent of the contract provisions is quite clear and – apart from variation instructions which are only a glint in the architect's eye – a competent contractor must have gained sufficient information from the contract documents to know what he has to do and from that work out what further information he will need and when he will need it. The position is elementary.

There is nothing in the contract to prevent a contractor from providing a complete schedule of information required and the dates when the items will be needed. This can even be done before work starts. It will certainly be useful to the architect but, despite the *Leach* decision will not usually (if ever) be a specific application for the whole of the contract period.

The programme in *Leach* was sufficient for the earlier operations shown on the programme but, as the judge emphasised

'it does not follow that the programme was a sufficiently specific application made at an appropriate time in relation to every item of information required, more particularly in light of the delays and the rearrangement of the programme for the work'.

There is, of course, and can be no requirement that application from the contractor is needed for variation instructions, and these are indeed dealt with separately by clauses 25.4.5.1 and 24.2.7. As regards other information, the contractor only gets the benefit of having the architect's breach dealt with contractually if he has specifically applied to the architect for it.

Legally, the situation is straightforward. In providing information necessary for the execution of the works and at the proper time, the architect is acting as agent for the employer who, as principal, is liable for his agent's breach of contract. Under JCT 80 terms the duty is to provide drawings and details 'as and when from time to time may be necessary'. The implication is that the information will be issued from time to time as the contract progresses at a time which is reasonable in all the circumstances.

Failure to do that entitles the contractor to damages at common law. But if he is to get the benefit of having the breach dealt with under clause 26, and so secure early reimbursement of his loss and expense, he must make a specific application as required by the contract and in accordance with its terms.

Chapter 8

Discharge and Determination

Inflation and fixed prices

Is galloping inflation a frustrating event which will bring a contract to an end? This is a question which is frequently asked by contractors working under fixed-price contracts and even, on occasion, by those working under contracts with a fluctuations clause.

It has long been my view that the answer to this question is a negative one under JCT terms because of the fluctuation provisions. In *Davis Contractors Ltd* v. *Fareham UDC* (1954), Lord Radcliffe emphasised that frustration can only occur

> 'whenever the law recognises that, without default of either party, a contractual obligation has become incapable of being performed because the circumstances in which performance is called for would render it a thing radically different from that which was undertaken by the contract.'

A Court of Appeal case of direct relevance is *Wates Ltd* v. *Greater London Council* (1983), which arose out of a contract for the construction of 1807 housing units at Hendon Aerodrome. The contract was in the GLC version of JCT 63, and included a fluctuations provision (clause 31) which, after the contract had run for some time, both parties agreed was not properly compensating Wates for the then current (1972) level of inflation.

A Supplemental Agreement was entered into in October 1972 to compromise various disputed claims and substitute an alternative type of fluctuations provision. This was on the basis of the Ministry's 'Housing Cost Yardstick' which had been regularly updated in the years before 1972. The Ministry changed its policy and no longer updated the yardstick, and this caused financial loss to the contractor.

Wates asked for £421 000 based on its interpretation of the Supplemental Agreement and, when the GLC refused to pay, treated that refusal as a wrongful repudiation of the contract and withdrew from site. The GLC contended that this was a repudiation of the contract and the dispute went to arbitration, and thence via the High Court to the Court of Appeal. The trial judge found that the GLC had not repudiated the contract and that the Supplemental Agreement had not been frustrated by inflation, as contended by Wates.

In the Court of Appeal, much of the judgment was concerned with the frustration point. The arbitrator had found in favour of the contractor, holding that the supplemental agreement was frustrated in that

'by February 1 1974 the circumstances in which the contract fell to be performed by Wates had become radically different from those contemplated by the parties at the date of the [agreement].'

In this he was wrong, said the Court of Appeal.

Although their lordships had considerable sympathy with Wates which suffered loss and damage, it was not possible to say that the terms of the Supplemental Agreement did not in law cover all the eventualities that occurred. The contractor merely got less than it hoped for. Despite an increase in the rate of inflation which it was accepted was unforeseen at the time of entering into the Supplemental Agreement and the Ministry's change of policy, this was not frustration.

Lord Justice Stephenson put it this way:

'Things may have turned out differently from what the parties contemplated in that inflation increased not at a trot or a canter, but at a gallop. But that difference in degree and tempo was not so radical a difference from the inflation contemplated and provided for as to frustrate the contract. It could only frustrate the contract if it were coupled with the (Ministry's) failure to keep up with it and provide for it by increasing the Housing Cost Yardstick or by some other method. And in fact the contract provided for it ... though not as effectively as Wates would have liked if they had contemplated it.'

Although *Wates Ltd* v. *Greater London Council* is a 'one-off' case, the principle laid down seems to apply to all fluctuation clauses. It does not deal with fixed-price contracts and the passage quoted from Lord Justice Stephenson's judgment lends some support to the view that raging inflation *might* conceivably frustrate the contract and so bring it to an end. In that event, of course, the contractor would be entitled to recover a reasonable sum.

That was, indeed, one of the points at issue in *Davis Contractors Ltd* v.

Fareham UDC itself, where the contractor agreed to build 78 dwellings for the council within a period of six months and for a fixed price. Through no fault of either party, there was an extreme scarcity of skilled labour and the work took 22 months to complete. The arbitrator found that the footing of the contract was so changed that it was frustrated and held that the contractor was entitled to a fair and reasonable price for the work it had done.

The House of Lords held that the claim must fail; the scarcity of labour had not frustrated the contract and the plaintiff was not released from its terms as regards price. The actual cost to the contractor of carrying out the work was in excess of £115 000, and the sum paid to the firm under the contract was £94 425. The contract was not terminated by operation of law because no fundamentally different situation had emerged. In effect, it seems that the events relied on must be fundamental enough to transform the job the contractor had undertaken into a job of a different kind, which the contract did not contemplate and to which it could not apply.

A careful reading of the *Davis* case, and especially the classic judgment of Lord Radcliffe, leads one to suggest that even in a fixed-price situation, inflation would not amount to a frustrating event. The arbitrator in that case took the view that if, without default on either side, the contract period was substantially extended, that rendered the fixed price unfair to the contractor and he ought not to be held to his original price. The House of Lords said that was bad law, and it is thought that the same principle applies to inflation.

Breach and damages

A breach of contract does not, of itself, bring the contract to an end. If, for example, the employer fails to give the contractor possession of the site in sufficient time for him to complete the works by the contractual date, this is a breach of contract. Indeed, it is a breach of one of the basic contractual obligations, entitling the contractor to accept the breach and treat the contract as at an end. Few contractors would wish to adopt that drastic course, but in any event the contractor is entitled to damages for any loss caused to him by the breach.

Damages are compensatory and not punitive, and case law has established the basic principles involved. The claimant is to be put in the same financial position as he would have been had the contract performed, so far as money can do this. Thus, if the contractor does throw up the contract for a breach on the employer's part, he is clearly entitled to claim the profit which he would have made had the job proceeded to completion: *Wraight Ltd* v. *P. H. and T. Holdings Ltd* (1968).

Damages are subject to the 'foreseeable test' laid down in the old case of *Hadley* v. *Baxendale* (1854). This means that the contractor can recover as damages only that part of his resultant loss as was reasonably foreseeable as liable to result from the breach. This is to be judged at the time the contract was entered into and not with hindsight, and this principle defeats many potential heads of claim.

Contrary to popular opinion, the amount of damages awarded bears no relationship to the contract price. A contractor or sub-contractor who performs badly or not at all may well find himself facing a claim for damages far in excess of his contract price.

This is illustrated by *Harbutt's Plasticine Ltd* v. *Wayne Tank and Pump Co. Ltd* (1970), where the defendants contracted to design, supply and erect equipment for storing the materials used in the manufacture of plasticine. They used plastic piping, surrounded by a heating element, as part of the system. This was unsuitable for its purpose. The defendants' employees started up the newly installed system and left it on overnight; over-heating occurred and the resultant fire destroyed the plaintiff's factory. The cost of the job was about £2500, but the defendants were held liable for damage totalling some £146 000. That case is merely cited as being factually illustrative, since its main point is no longer good law.

There is another basic principle which is often overlooked. The claimant must have taken reasonable steps to mitigate his loss and, in the words of Lord Haldane, this rule 'debars [the claimant] from claiming any part of the damage which is due to his neglect to take such steps'. In a building contract context, for example, the contractor must re-deploy his men and other resources elsewhere: he cannot merely sit back and let the losses mount up.

Today, the courts appear to adopt a more flexible approach to the measure or amount of damages to be awarded in particular breach situations. In particular, the so-called rule in *The Edison* (1933), is being more and more disregarded. In that case Lord Wright said that any loss due to the impecuniosity of the claimants was not recoverable.

Although that was a decision of the House of Lords, it was not followed in either *Dodd Properties (Kent) Ltd* v. *Canterbury City Council* (1980), or in *Perry* v. *Sydney Phillips and Son* (1982) where it was described as 'not of general application.'

Thus, if a contractor suffers cash flow problems and incurs finance charges as a direct consequence of the breach, in principle these would now be recoverable as special damages: *Holbeach Plant Hire Ltd* v. *Anglian Water Authority* (1988).

Difficult problems arise where one party throws up the contract before the time fixed for performance – a situation called by lawyers 'anticipatory breach'. The innocent party has an option: either he may accept the breach and sue at once *or* he may hold the defendant to the contract and await the date of performance before doing anything.

The authority for this statement is the decision of the House of Lords in *White and Carter (Councils) Ltd* v. *McGregor* (1961). The claimants there supplied litter-bins to local authorities and agreed with the defendants to advertise their business on the bins. It was a three-year contract. The defendant repudiated the contract on the day it was made. The claimants still went ahead and took no steps to minimise their loss. The contract provided for the full contract price to become due if there was default in any one instalment, and the claimants sued for that sum without waiting for the three years to elapse.

The House of Lords held that they were entitled to succeed:

'There is no duty laid on a party to a subsisting contract to vary it at the behest of the other party so as to deprive himself of the benefit given by the contract. To hold otherwise would be to introduce a novel equitable doctrine that a party was not to be held to his contract unless the court in a given instance thought it reasonable so do to.'

The decision would, of course, have been different if the claimants had deliberately inflated their loss, but it stands as a warning to those who enter into contracts and then breach them before the time fixed for performance. A contractor who has taken on an unprofitable contract would be caught by the decision if he threw the contract up.

Breaking a contract can be a costly business, as many a contract breaker has found out. If a contractor fails to start the work, the damages awarded to the employer might be the difference between his and the next satisfactory tender, together with the employer's loss of ordinary business profits, assuming that the works are plainly for a profit-making concern.

However, if the contractor had undertaken to build a supermarket in time for the Christmas trade and broke his contract in the same circumstances, the measure of damages would be the profits lost in the period before another contractor could start work and the cost of engaged but non-working staff together with any increased cost of building.

Wasted expenditure

Damages for breach of contract are intended to put the plaintiff as far as possible in the position which he would have been had the contract been performed. He is not entitled to be put in a better position, and it is for the plaintiff to prove both the fact and the amount of the damages. These elementary principles apply equally to claims for loss and/or expense (or its equivalent) under the industry's standard forms.

The case of *CCC Films (London) Ltd* v. *Impact Quadrant Films Ltd* (1984) suggests that, in appropriate circumstances, instead of claiming loss of profits as a head of damage, the plaintiff can claim wasted expenditure instead. He has an unfettered choice under either head and is not restricted to claiming for wasted expenditure only where he establishes by evidence that he could not prove loss of profits, or that his loss of profits would have been small.

This decision may be of limited application in a claims situation under the standard forms of contract, but the judgment needs to be read very carefully for its facts. Mr Justice Hutchinson was prepared to assume in the plaintiff's favour that had the contract in question been performed, the plaintiff's profits would at least have come up to the $12 000 expenditure which was claimed, and in allowing the claim for that amount he reviewed the law very carefully.

In *Anglia Television Ltd* v. *Reed* (1971), a decision of the Court of Appeal, an actor repudiated his contract with the result that the television play had to be abandoned. The plaintiffs claimed £2750 wasted expenditure as damages for breach of contract, of which the greater part had been laid out by them before the contract was entered into. The issue was whether both the pre- and post-contractual expenditure could be recovered, and no argument was put forward as to whether, had the production gone ahead, it would have generated sufficient income to cover the outlay.

Lord Denning MR allowed the plaintiff's claim for wasted expenditure. He said that in such a case the plaintiff has an option. He can claim either loss of profit or wasted expenditure. He cannot claim both, but if he claims wasted expenditure

> 'he is not limited to the expenditure incurred *after* the contract was concluded. He can claim also the expenditure *before* the contract, provided it was such as would reasonably be in the contemplation of the parties as likely to be wasted if the contract was broken.'

This important point – for which there was earlier authority – was seized on by Mr Justice Hutchison, who added a gloss of his own. The claimant may always frame his claim in the alternative way if he chooses, because:

> 'to hold that there had to be evidence of the impossibility of making profits might in many cases saddle the plaintiff with just the sort of difficulties of proof that this alternative measure is designed to avoid.'

However, as the judge pointed out, a claim for wasted expenditure cannot succeed in a case where, even had the contract not been broken by the defendant, the returns earned by the plaintiff's exploitation of the contractual right would not have been sufficient to recoup the expenditure. The authority for that is the Court of Appeal decision in *C. & P. Haulage* v. *Middleton* (1983), where the plaintiff sought to maintain a claim for the cost of work to premises from which he was later unlawfully evicted. The evidence established that the plaintiff was actually better off as a result of being evicted than he would have been were he permitted to remain until the time when he could lawfully have been required to leave.

So, in the case of breach of contract by the employer, where the contractor is losing money hand over fist and is working at a loss rather than a profit, there is no benefit to be gained in framing the claim as one for wasted expenditure rather than for loss of profit. The significance of *CCC Films Ltd* v. *Impact Films Ltd* is that, if correct, it is not for the plaintiff to prove that his profits would at least have matched the wasted expenditure claimed, but for the defendant to prove otherwise.

Mr Justice Hutchison was careful to distinguish between the term 'loss of profit' and the term 'recovery of expenditure.' He said:

> 'When Lord Denning MR speaks in *Anglia Television Ltd* v. *Reed* of the plaintiffs not having suffered loss of profits or of it being impossible for them to prove what their profits would have been, he is referring ... to profits after recoupment of expenditure i.e. net profits.'

He pointed out that, often, the difficulties of proving loss of profits

> 'would clearly have been enormous and it is hard to envisage how the plaintiffs in the present case could in practice have proved a claim based on loss of profits,'

even though the proof is on the balance of probabilities.

The judgment in *CCC Films Ltd* answers the crucial question of where the onus of proof lies in relation to whether or not the exploitation of the subject matter of the contract would or would not have recouped the expenditure by saying that it lies on the defendant. Interestingly, there appears to be no earlier English judgment directly in point and so Mr Justice Hutchison turned to an American case *Albert & Son* v. *Armstrong Rubber Co.* (1949), where a claim for damages for breach of contract – in relation to the sale of some machines designed to recondition old rubber – was advanced on the wasted expenditure rather than the loss of profits basis. In such a case, the judge said, the burden is on the defendant to prove that full performance of the contract would have resulted in a net loss.

Mr Justice Hutchison felt that this rule was just:

'It appears to me to be eminently fair in such cases, where the plaintiff has by the defendant's breach, been prevented from exploiting the right contracted for and, therefore, putting to the test the question of whether he would have recouped his expenditure, the general rule as to the onus of proof should be modified in this manner.'

The 'wasted expenditure' approach may be of *some* assistance in some claims situations, but not, it seems, where the claimant can turn to an equally profitable contract elsewhere without extra cost.

Battle for title

The validity of a materials supplier's retention of title clause was at issue in the interesting case of *W. Hanson (Harrow) Ltd* v. *Rapid Civil Engineering Ltd and Usborne Developments Ltd* (1987), where the main contractor had gone into receivership, leaving the timber supplier, Hanson, unpaid.

Hanson had traded with Rapid since 1979 on terms and conditions which included a retention of title clause. This provided:

'(a) The property in the goods shall not pass to you until payment in full of the price to us.
(b) The above condition may be waived at our discretion where goods or any part of them have been incorporated in building or constructional works.'

Rapid and Hanson carried on business on a running account.

Hanson supplied timber for three sites which Usborne was developing and where Rapid was the contractor. By July 1984, Rapid was becoming 'slower and slower' in paying Hanson, and Usborne, the developer, began to have cash-flow problems. Usborne decided to make more frequent payments under the contracts based on valuations of on-site materials as well as work done. Two valuations were made, but in the event, the sums involved were not paid.

On 15 August 1984 Rapid went into receivership. The receivers allowed Hanson to take an inventory of the materials on site but they refused to make payment or return the goods. On 5 September 1984 Hanson notified Usborne of the retention of title clause and, in the absence of an assurance that the goods were not being used, they issued a writ for conversion of the materials worth £6768.97.

Section 25(1) of the Sale of Goods Act 1979 enables a buyer of goods to transfer title to them by way of 'sale or disposition' to 'any person receiving the same in good faith and without notice of any right of the original seller in respect of the goods.' His Honour Judge John Davies QC held that there had been no delivery or transfer of the loose materials on site by the contractors to the employers. The building contracts were lump sum contracts. One of them, whilst providing for stage payments on account of a percentage of the contract price, made no provision for either the stages or percentage referred to. The other contracts provided for monthly payments of 97% of the value of work executed and materials on site. But they also said that property in the goods would not pass to the employer until the stage payments were made. Thus there was a prior agreement to sell to the contractor. There was no prior agreement to sell to the employer and unless the contractor had acquired title from the supplier, he had no title to pass to the employer, even if the contract provided that it would pass on payment.

The first contract made no provision for interim payments or the passing of property in goods on site. The judge accepted that there was an arrangement for interim payments but said:

> 'Valuation of work and materials for that purpose does not usually connote the purchase by the employer of site materials so valued. The assessment is merely a convenient means of determining the amount which should, in fairness to the contractor, be advanced to him from time to time against the contract sum.'

Once again section 25(1) was inapplicable.

The matter did not end there, because the Usborne-Rapid contracts contained a 'bankruptcy clause'. This provided that if the contractor

(Rapid) committed an act of bankruptcy, or entered into involuntary liquidation, the employer (Usborne) could determine the contractor's employment by written notice, and then could use 'any materials of the contractor' on the site for the purpose of continuing the work. Under the clause, the employer was not bound to pay any more money to the contractor until the work ought reasonably to have been completed. When that happened, the contractor was to be credited with the value of any work he had done, including the value of 'any materials of the contractor'.

This is a common form provision in most construction contracts, and may usefully be compared with JCT 80, clause 27.4, which sets out the rights of the parties following employer determination. Not surprisingly, Judge John Davies QC was unhesitatingly of the view that Hanson's materials were not those of the contractor. Hanson had retained title in the goods. Although a provision of this kind is binding between the contracting parties, it can only extend to materials which belong to the contractor in law. A 'bankruptcy clause' cannot confer rights on the employer in respect of goods which belong to someone else, as in this case where there was a valid retention of title clause. The 'bankruptcy clause' is defeated by the rights of third parties: see, for example, the well-known case of *Dawber Williamson (Roofing) Ltd* v. *Humberside County Council* (1979).

The receivers had not, in fact, laid any claim to the goods and in any case the learned judge was not satisfied that there had been any written notice of determination as required by the relevant contract provision, or that the circumstances justified Usborne relying on the clause. He continued:

'On August 22 1984 formal notice was given to Usborne by the receivers that they were treating all three contracts as having been discharged by Usborne's repudiatory breach of them – by their failure to pay in accordance with them. As from that time there were no effective contracts for Usborne to determine: they had not purported to determine them before that time.'

Accordingly, the learned judge ruled in favour of the supplier.

In some ways it is surprising that this matter should have proceeded to the High Court, having regard not only to the comparatively small amount involved but also to what, with hindsight, appears to be very elementary law. But the case demonstrates the usefulness of retention of title clauses from the supplier's point of view and reminds architects of the need to check sales invoices before certifying for payment of unfixed goods and materials. An employer who, in effect, had to pay twice, could well sue the architect for professional negligence if he had not done so.

Ultimate remedy

Clause 28 of JCT 80 sets out the grounds on which the contractor can bring an end to his employment under the contract. The contract itself remains in being; it is only the contractor's employment under it which terminates and, indeed, the clause sets out the rights and duties of the parties when the power it confers is invoked. One of the grounds on which the contractor can so end his employment is if the employer does not honour certificates in due time.

Clause 28.2.2 sets out some of the consequences of determination and the contractor's entitlement to payment. This includes 'any direct loss or damage caused to the contractor or any nominated sub-contractor by the determination', and as John Parris remarks in *The Standard Form of Building Contract*, 1982, p. 161

> 'the deadly result of (this) is that the employer has to pay the contractor and every nominated sub-contractor the whole of the profits they would have made if the contract had been carried to completion.'

It may be, however, that even though the contractor has validly exercised his right to end his employment, he has also done defective work, and this aspect of valuation was discussed by the High Court in *Lintest Builders Ltd* v. *Roberts* (1979), which gives a sensible answer to the problem.

Mr Roberts employed Lintest to convert an oast house into a dwelling. The contract was in the JCT 63 standard form, without quantities. Mr Roberts failed to honour certificates within the stipulated time and Lintest determined its employment. The disputes between the parties went to arbitration, and there was an allegation that some of Lintest's work was defective.

In valuing the work done, the architect made a gross valuation as if all the work had been properly done and then deducted the sum at which he valued his snag list of defects and omissions. Lintest argued that the correct approach was to value the work which had been properly done and to make no deduction for work which had not been properly carried out. The arbitrator submitted this simple question for the decision of the court:

'Whether the reasonable cost of the necessary remedial works is to be taken into account when calculating sums due to (the contractor) pursuant to and in accordance with'

the determination provision?

Mr Justice O'Connor answered that question in the affirmative. The reasonable cost of the necessary remedial works is to be taken into account when calculating the sums due to the contractor because the employer has a right to have the bad work put right or corrected. This right accrues as the contract proceeds, and the judge declined to follow some broad generalisations of Lord Diplock in *P. and M. Kaye Ltd* v. *Hosier and Dickinson Ltd*, (1972), which suggested a contrary conclusion.

The judge took the view that if there are a number of defects which are unremedied at the time when the contractor properly brings his employment to an end, then there is an accrued right in the employer to have these defects put right.

'How that accrued right is satisfied is a different matter ... the contract is not determined. The contract provides for the various ways in which that accrued right is to be dealt with. Sometimes, if the defects are latent or hidden or do not appear until the defect period, the contract provides for what is to be done ... it seems to me that as to how the valuation of the accrued right to have the defects cured is done, is a matter of fact to be dealt with by the parties and, failing them, in arbitration ...'

Clause 28 is much disliked by employers, and the various legal commentators are critical of the provision largely because it confers a right of determination on the contractor for matters which are not necessarily breaches of contract at common law. Dr Parris refers to the provision as being 'extraordinarily onerous' on employers and suggests that private employers 'would be well advised to delete the whole clause and allow the contractor to rely on his common law rights'.

I have never understood these criticisms and would advise contractors should resist any attempt to delete clause 28 as it has been inserted to mitigate the difficulties and uncertainties of establishing a claim for breach of contract at common law. In general, for the contractor to have a remedy at common law there would have to be a breach by the employer going to the root of the contract. Failure to pay on time is not such a breach, although non-payment of more than one certified sum may well be: see *D. R. Bradley (Cable Jointing) Ltd* v. *Jefco Mechanical Services Ltd* (1988).

Thus the provisions of clause 28.1.1 (as invoked in the *Lintest* case) are a very necessary protection for the contractor and provide him with an effective sanction. Why it should be thought that clause 28 is onerous to employers is difficult to see; it is a simple and extraordinarily effective remedy, if properly used.

True, some of the grounds listed are not 'defaults' on the employer's part; but taken as a whole clause 28 is a sensible and business-like provision covering those matters which are of common concern to all contractors.

Those who criticise the clause apparently overlook the fact that it has been freely negotiated within the JCT – which is a body representative of both employer and contractor interests – and given that the contractor should be entitled to determine his employment in carefully prescribed circumstances, there is nothing wrong with the provision.

Moreover, there is an important protection for the employer, because the exercise of the power of determination is subject to this important proviso: the notice of determination 'shall not be given unreasonably or vexatiously'. The two recent cases of *J. M. Hill & Sons Ltd* v. *London Borough of Camden* (1980) and *John Jarvis Ltd* v. *Rockdale Housing Association Ltd* (1986) give guidance on the meaning of these words. 'Unreasonably' here is a general term which can include anything which can be objectively judged to be unreasonable, while 'vexatiously' connotes an ulterior motive to oppress or annoy.

The consequences of a wrongful determination are very serious from the contractor's point of view, and in my experience contractors have been very loathe to invoke the determination provisions. The payments and remedies provided for by clause 28 are additional to his common law rights, but the strictures of so many commentators are not supported by the facts.

A very strong case can be made for the inclusions of similar clauses in all standard-form contracts, not least for unjustified failure by employers to honour certificates in due time.

Insolvency solution

The annual Government statistics show that the number of insolvencies is increasing, leading to more and more companies going into liquidation. The standard contract forms protect the employer when a contracting company goes into liquidation, and this contractual protection is necessary because the liquidator's duty is to the body of creditors as a whole, and under the general law the employer is simply an ordinary unsecured creditor.

Under JCT 80 clause 27.2 provides that on a contractor's insolvency 'the employment of the contractor under this contract shall be forthwith automatically determined ...' there is a provision for reinstatement by agreement between the parties and the liquidator, but this seems unimportant and indeed is so self-evident as not worthy of being expressed in the contract.

Clause 27.4 contains the provisions which govern the relations between the parties after determination. The contract continues to govern the relationship of the employer and the contractor: it is simply that the contractor no longer has any right to continue his work, since his employment has been determined.

The employer's rights under clause 27.4. are:

(1) he may engage another contractor to carry out and complete the works;
(2) he, and the substituted contractor, may enter upon the works and use the insolvent contractor's own temporary buildings, plant, tools, equipment, goods, materials, etc, 'intended for, delivered to and placed on or adjacent to the works';
(3) he may purchase all materials and goods necessary for carrying out and completing the works. (He would have this right in any event).

   From the employer's point of view, after his quantity surveyor has assessed the situation as soon as determination has occurred, it is vital to make arrangements for the completion of the contract works. This must be done as economically as possible. In lawyer's language, the employer must 'mitigate his loss', and this normally means going out to competitive tender. Emergency works will, inevitably, have to be done at daywork rates.
(4) The employer's former right to pay direct any supplier or sub-contractor is now excluded if the reason for the contractor's employment being determined is insolvency. However, so far as nominated sub-contractors are concerned, clause 35.13.5 confers a right of direct payment in the circumstances there set out and, indeed, an obligation to pay where Agreement NSC/2 is in operation, but that obligation terminates on the contractor's insolvency.

There are many problems involved in the ownership of materials on site. The JCT contract transfers ownership when the materials are paid for in a certificate (see clause 16), and clause 27.4.1 expressly provides for the employer to use any materials on site in completion of the work, but that provision applies only to materials actually owned by the contractor. This point was emphasised in *Dawber Williamson Roofing Ltd* v. *Humberside County Council* (1979), where a main contractor went into liquidation, and the employer claimed as his property some slates delivered on site by a sub-contractor, the value of which had been included and paid in a certificate to the main contractor. The equivalent of JCT 80, clause 16, was in operation.

The High Court held that the clause could only transfer title in the slates if the main contractor had a good title to them, which he had not. Consequently, the plaintiff sub-contractors were entitled to judgment against the employers for the value of the slates on the basis that the employer had been guilty of the tort of conversion. This problem can, of course, be avoided by an appropriate sub-contract term – and no doubt this is a point being checked by quantity surveyors now before they include the value of the relevant materials in any certificate. It is not safe to rely on the Joint Contract Tribunal's amendments introduced in an attempt to reverse the *Dawber Williamson* case: see Powell-Smith, *Contractor's Guide to the JCT Standard Form of Building Contract (JCT 80)*, 2nd edition, 1988, pp. 75–76.

The remaining parts of clause 27 deal with the contractor's obligations when his employment has been determined and payments that must eventually be made.

Clause 27.4.3 requires the contractor to remove his temporary buildings, plant, tools, equipment, etc. from the site 'as and when required in writing by the architect to do so'. Default powers are conferred on the employer who may 'remove and sell any such property of the Contractor' if the latter has not complied with the architect's instruction within a reasonable time. The net proceeds of sale are held to the credit of the contractor and it is to be noted that the employer is expressly exempted from any liability for any loss or damage to the contractor's property.

Clearly, the clause does not extend to goods, etc. hired to the contractor, and the bulk of plant is likely to be hired or hire-purchased. If the plant is in the latter category, the employer has no rights in it at all.

Money matters are covered by clause 27.4.4. The contractor must 'allow or pay to the employer ... the amount of any direct loss and/or damage caused to the employer by the determination'. The employer's claim is set off against any sums due to the contractor and the employer has a right to suspend payment of any moneys due until after completion of the works. If nothing is due to the contractor, the employer must prove it as an unsecured creditor in the liquidation.

Although the point is not free from doubt, it is suggested that any liquidated damages due can be taken into account in the calculation under clause 27.4.4, and, since the phrase 'direct loss and/or damages' is the equivalent of the common law position, the employer's claim can be quite extensive. The phrase will include additional professional fees incurred as a direct result of the contractor's employment being determined and, by analogy with the contractor's position, in 'direct loss and/or expense' claims, interest on additional capital employed.

Chapter 9

Disputes Settlement

The powers of the arbitrator

The powers of construction industry arbitrators were strengthened by the Court of Appeal judgments in *Ashville Investments Ltd* v. *Elmer Contractors Ltd* (1987), which reconciled earlier conflicting authorities.

Elmer contracted with the building owners for the construction of six warehouse units at Wokingham, Berkshire. The contract between them was in JCT 63 form and was dated 22 December 1982. Negotiations for the work began in April 1982, and Elmer tendered on the basis of a specification and drawings.

At the end of a pre-contract meeting on 22 December 1982, an amended JCT 63 contract was signed and sealed. There were substantial differences between the specification and drawings on which Elmer tendered and those which were initialled at the pre-contract meeting and incorporated into the contract. Elmer contended that it was unaware of these differences until well after the works had begun. Ashville's evidence was that the differences were drawn to Elmer's attention at the meeting and, in spite of the cost implications, it had decided not to change its tender price of £715 000.

Clause 1(1) of the contract had been amended so as to oblige Elmer to 'carry out and complete *the design* and the works shown upon the contract drawings, etc'. This matter was also discussed at the meeting. The parties again disputed what was agreed, though the architect's minute read that clause 1(1) was 'a reference to heating, plumbing and electrical installations'. Ashville argued that the minute was not contractual and that Elmer's obligations as to design were not limited.

The contract overran and was not practically completed until 8 December 1983 – several months late. Elmer claimed more than £150 000 in respect of variations and disruption, and the dispute was referred to arbitration. Part of Elmer's case was that the formal contract did not represent the true agreement between the parties.

It claimed to have it rectified on the grounds of mistake. It also sought damages for innocent misrepresentation or negligent mis-statement.

The dispute came before the courts because, so it was said, the arbitrator had no power under the arbitration agreement (clause 35 of JCT 63) to order rectification or award damages for misrepresentation.

An allegation of technical fraud had also been made – though this was later withdrawn – and the court was also asked to order that the dispute be litigated rather than arbitrated: Arbitration Act 1950, section 24(2).

Despite earlier decisions which could have led to another conclusion, a very strong Court of Appeal ruled in favour of the contractors, and held that the arbitrator was empowered to grant relief in respect of operative mistake, misrepresentation or negligent misstatement and to order rectification of the contract. This was the plain meaning to be given to the very wide terms of the arbitration clause.

Lord Justice May made a number of valuable points:

- Whether a particular dispute falls within any arbitration clause depends primarily upon the interpretation of the clause itself.
- If a dispute is within the arbitrator's jurisdiction, he is bound to give the necessary remedy to the successful party, provided the clause does not debar him from doing so.
- Under JCT terms the arbitrator is empowered to grant wider relief than may be available in the courts. He can 'open up, review or revise' the architect's opinions, certificates and so on.
- There was no reason in principle why an arbitrator could not order rectification. Earlier cases on the topic could be explained as matters of interpretation of the particular clause being dissected.
- The Court of Appeal also pointed out in *Northern Regional Health Authority* v. *Derek Crouch (Construction) Ltd* (1984) that not only are the powers and duties of the architect, the agent of the building owner, under the JCT forms wide indeed, but it is quite clear that the supervisory powers of the arbitrator are and are intended to be wider still. Lord Justice May said:

> 'I have no doubt that it would be wrong to restrict those powers but desirable in law to hold that they are as wide as the actual wording of the arbitration clause permits'.

He – and the other members of the Court of Appeal – were in no doubt about the matter, holding that there was no binding authority to the contrary.

Elmer had abandoned any allegation of fraudulent misrepresentation, and following the views expressed by Lord Wilberforce in *Camilla Cotton* v. *Granadex* (1976), the learned Lord Justice held that the court's jurisdiction under section 24(2) of the 1950 Act 'can only be exercised where a concrete and specific issue of fraud has been raised'.

Alternatively, it was a proper exercise of the judge's discretion to refuse to revoke the arbitrator's authority.

Ashville Investments Ltd v. *Elmer Contractors Ltd* emphasises just how wide are the powers of an arbitrator acting under JCT contract terms; and the JCT has now given the arbitrator *specific* power to rectify the contract – presumably to avoid arguments.

Lord Justice Balcombe, after quoting clause 35(1), emphasised that there was

> 'no reason why [its] words should be limited only to matters that arose after the making of the contract, and should not include matters which happened before the making of the contract, provided that they are connected [with it].'

The judgments must be cold comfort to those who opposed the *Crouch* principle and hoped for a decision which would limit the effect of that case.

As Lord Justice Bingham pointed out, there is no class of contract in which the content of the fundamental terms (price, contract period, variations, etc.) is potentially as fluid as in the standard construction industry forms. Authority to determine the work done, the contract period and the price to be paid is vested in a professional agent: the architect or engineer.

The width of his powers means that they must be subject to review and control, and hence the contract defines the arbitrator's jurisdiction in very wide terms. The reference to 'any dispute or difference' contains no limitation of any kind, except by reference to the subject matter – the contract. The test to be applied is a simple one: does the dispute or difference arise out of the contract or not?

Maker or minder?

Is the architect a contract maker or a contract administrator? This was one of the points at issue in *Partington & Son (Builders) Ltd* v. *Tameside Metropolitan Borough Council* (1985), which is one of the growing number of cases reaching the courts as a consequence of the Court of Appeal ruling in *Northern Regional Health Authority* v. *Derek Crouch (Construction) Ltd*

(1984), about the jurisdiction of the courts to entertain actions relating to JCT contracts containing the standard arbitration clause.

Partington issued a writ for £51 000 which the firm claimed was money due to them under an interim certificate allegedly issued by the architect, though the council did not admit that the document was a certificate or, if it was, it claimed that it was not issued because the council witheld its issue and was entitled to do so. There were other disputes between the parties about liquidated damages, extensions of time, loss and expense and so on.

The upshot of the initial litigation was that Partington was refused summary judgment on the writ in 1981 and the litigation proceeded at a leisurely pace. The parties had agreed that, subject to the court having jurisdiction to hear it, they wished the action to proceed on the basis that the court would have all the powers of an arbitrator under JCT terms – including power to open up, review and revise the architect's certificates, opinions, and decisions.

His Honour Judge John Davies QC had, therefore, to consider the architect's role under JCT contracts, and in particular whether the exercise by the architect of his powers results in the creation of new contractual rights and obligations between the parties, or whether his role is merely a functional one concerned with the execution and administration of the contract. Similar questions arose in relation to the arbitrator's role in relation to certificates, opinions and so on.

The learned judge gave a careful analysis of the scope of the architect's functions. Firstly, he acts as the employer's agent in issuing instructions to the contractor, including variation instructions. These may certainly modify the contractual rights and obligations of employer and contractor, but since the architect is acting as the employer's agent, in the judge's view they do not fall within the scope of the arbitration clause.

The architect is also 'the impartial administrator of the contract ... to ensure its orderly and efficient progress in accordance with its terms'. He has other specific powers and duties, some of which are qualified by the expression 'when in the opinion of the architect'.

The judge attached no significance to this, holding that the exercise of those types of power did not involve contract making. The phrase, he said:

> 'merely expresses the material time for coming to a decision on the facts. The occasions when the architect's discretion comes into play are few ... the exercise of that discretion is so circumscribed by the terms of ... the contract as to emasculate the element of discretion virtually to the point of extinction.'

His Honour was referring in particular to JCT 63, clause 30(2A) dealing with the value of off-site goods and materials.

As regards the architect's powers as certifier, and especially in relation to the issue of the final certificate, the judge was equally emphatic. The issue of certificates is merely part and parcel of the machinery of implementing the contract in accordance with its agreed terms. It is not part of some contract-modifying function vested in the architect.

Judge Davies concluded:

'The architect in his capacity as administrator of the contract, rather than agent of the employer, has no power to modify or supplement the terms of the contract agreed between the parties, his functions ... are merely to carry it out. When they are the subject of complaint it is on the ground that they do not properly reflect what the parties agreed.'

With great respect, this is undoubtedly correct. Architects do not possess any dispensing or modifying powers; in contractual terms the architect is a mere administrator, though charged in some cases with a duty to act impartially.

The judge went on to consider the arbitrator's role in light of the wording of the arbitration clause. He similarly concluded that the arbitrator is not a contract maker or modifier. Since the architect's decisions do not become terms of the contract, the arbitrator merely has to consider whether or not the architect acted in accordance with the terms of the contract.

The validity of the contrary argument depends on giving a contractual effect to the architect's decisions. 'Such an attribution is, in my view, both unnecessary and unfounded,' he remarked. He went on to say

'there is no compelling reason to construe the "opening up" provision (in clause 35 of JCT 63) as being anything other than a clarifying indication to the arbitrator that while he is bound by the Final Certificate under clause 30(7), he is not similarly fettered in the case of other certificates and decisions of the architect.'

Interestingly, he concluded that:

'the powers of an arbitrator under clause 35 are no different from those which would be possessed by a judge if a court were to be seised of the dispute', thus doubting what was said in *Crouch*.

This is not what has been generally understood to be the position in light of the observations of the Court of Appeal in *Crouch*. Evidently, the learned judge regarded *Crouch* as to some extent a policy decision. He took the view that the backlog of cases before the courts does not justify 'the elimination of the Court's discretion to disregard the arbitration clause when there are compelling reasons for doing so'. However, his views have not been followed by his brethren.

In *Partington* both parties wanted the court to exercise the arbitrator's powers and did not wish to arbitrate and obviously the judge is hoping for direct guidance from the Court of Appeal on the exact scope of *Crouch*. Other official referees have taken a different view to His Honour, e.g. the late Judge Smout QC in *Oram Builders Ltd* v. *Pemberton* (1985) and His Honour Judge John Newey QC in several cases. Accepting as we must, that the decision in *Crouch* is correct, the court cannot acquire jurisdiction that it never had merely by the parties agreeing to that course, or by agreeing a contract term providing that the court shall have wider powers. Of course the parties could concur in appointing the trial judge as arbitrator, subject to section 4 of the Administration of Justice Act 1970: see Parris, *Arbitration*, pp. 69–70.

Be quick to appoint an arbitrator

The meaning of one of the more obscure provisions of JCT 80 was the subject of litigation in the unreported but important recent case of *Emson Contractors Ltd* v. *Protea Estates Ltd* (1987). When is an arbitration 'commenced' for the purposes of clause 30.9 of JCT 80?

The clause says the final certificate is only to be 'conclusive evidence that where the quality of materials or the standard of workmanship are to the reasonable satisfaction of the architect'. The certificate is also

'conclusive evidence that any necessary effect has been given to all the terms of this Contract which require that any amount is to be added to or deducted from the Contract Sum or an adjustment is to be made of the Contract Sum ...'

Where arbitration proceedings are started by either party within 14 days from the issue of the final certificate, this certificate is conclusive evidence as defined, save in respect of matters to which the proceedings relate.

The contract in this case was for the erection of 16 nursery industrial units and external works at Letchworth. It was in JCT 80 terms and was dated 23 October 1984.

A dispute soon arose about whether certain foundation and sub-structure works made necessary by alleged unforeseeable ground conditions were a variation within clause 13. If so, this would entitle the contractor, Emson, to additional payment. Protea and the architect contended, however, that these works were within the contract.

There were other disputes about direct loss, expenses and extensions of time.

Emson produced its final account on 23 December 1985. This was sent to the quantity surveyor several months later. Emson and the architect corresponded about the dispute and on 29 July 1986, the question of a meeting was broached in a telephone conversation between Emson and the architect.

The quantity surveyor produced its detailed account on 2 September – £110 648 less than Emson's account.

Meanwhile, the architect had written to say that Protea was prepared to have a meeting only after the final certificate was issued. The certificate had to be issued under clause 30.8 by 2 September. However, the architect was prepared to delay the issue of the certificate until 16 September if he got formal confirmation for a meeting from Emson.

Emson agreed to the meeting but said that if the final certificate was issued while there were outstanding matters 'we would have little alternative than to dispute the issue in the manner set out in the contract.'

The meeting took place on 12 September. The quantity surveyor, however, quite properly refused to discuss the claim for unforeseen ground conditions. Since the architect said there was no variation, the quantity surveyor could not make a valuation of it.

Emson's remedy was to go to arbitration. The final certificate was issued on 16 September.

According to Emson's evidence, the final certificate was misfiled by a clerk who thought it was another interim certificate.

Part of it was obscured by a compliments slip and Emson testified that it was not until much later that the certificate's existence and significance became known. It was certainly before 6 January 1987, when Emson wrote to the architect about it.

Emson applied to the court under section 27 of the Arbitration Act 1950 to extend time for giving notice of arbitration on the ground that 'undue hardship would otherwise be caused' to Emson. The contractor also asked the court to rule the final certificate invalid because it contained 'technical errors'. Emson was described incorrectly and the contract date was shown as 16 October 1984, rather than 23 October.

His Honour Judge James Fox-Andrews QC ruled against the contractor on both counts. He concluded that:

'an arbitration commences under clause 30.9.3 when one party invites agreement to a named person being arbitrator or one party makes a written request to the other to concur in the appointment of an arbitrator, whichever is the earlier.'

While he accepted that Emson's ignorance of its position was a mitigating factor in respect of the delay in commencing an arbitration, that mitigating factor ceased on 6 January 1987. It was not until six weeks later that Emson tried to get an arbitrator appointed. It was then too late. The delay in appointing an arbitrator was 20 weeks after the issue of the final certificate. There was no explanation or mitigating circumstances for six of those weeks.

The judge also said the JCT contract is industry-negotiated and

'it seems reasonable to assume that the fact that there is a fundamental difference between 14 days and ordinary limitation periods is recognised and accepted by the' [JCT as desirable].

He thought the delay was significant and said 'that the loss of the opportunity of arbitration proceedings necessarily involves a great deal of hardship on Emson'.

From 6 January 1987, it must have been a conscious decision to disregard the arbitration proceedings, he said. There was no fault or misleading by Protea. 'To attempt to get negotiations going cannot be an excuse or mitigation' for the delay.

The technical objections to the final certificate were equally irrelevant. Emson was not misled by the certificate and the errors did not invalidate it.

The lesson is to get arbitration going sooner rather than later. Otherwise, contractors will be the losers.

Multi-partite problems

One of the main deficiencies in the Arbitration Acts 1950 to 1979 is that no provision is made for arbitrations to be heard together in the same way that actions may be tried together in court. This is of great importance in construction industry arbitrations where related disputes arise between main contractor and employer and between the main contractor and a sub-contractor.

The practical answer is to have a suitably comprehensive arbitration clause in each of the relevant main and sub-contracts. This the JCT main contracts and related sub-contracts attempt to provide. For various reasons, one of the disputing parties may wish to challenge the arbitrator's jurisdiction, and in recent years the problem has been before the courts on several occasions.

An excellent instance is *Multi-Construction (Southern) Ltd* v. *Stent Foundations Ltd* (1988), where the main contract was in JCT 63 form and the domestic piling sub-contract was in the FASS/NFBTE blue form, July 1971 edition, as revised in July 1978. The blue form arbitration agreement is contained in clause 27. It has an important proviso:

> 'Provided that if the dispute between the contractor and the sub-contractor is substantially the same as *a* matter which is *a* dispute or difference between the contractor and the employer under the main contract the contractor and sub-contractor hereby agree that such dispute or difference shall be referred to the arbitrator appointed or to be appointed ... under the main contract ...'

In the *Multi-Construction* case disputes arose as to whether a sub-contractor, Stent, had satisfactorily completed piling works, and with regard to its entitlement to payment. Multi failed to pay Stent sums certified under the main contract for piling work, and made claims against it for loss and/or expense. Shortly afterwards, the employer (Servite) purported to determine the main contract.

Stent asked Multi to concur in the appointment of an arbitrator on 30 October 1986. On 2 December 1986 Multi served notice of arbitration on Servite under the main contract. The arbitrator was agreed on 19 December. With the main contract arbitration notice, Multi wrote referring to the sub-contract dispute in general terms and invoked the proviso to clause 27 of the blue form asking that the same arbitrator determine both disputes.

Because of misunderstandings between the parties (for which nobody was to blame) the arbitrator's jurisdiction was disputed and the matter came before the court. It was argued that for the right of joinder to be exercised under the clause 27 proviso, at the date of Stent's request to Multi to concur in the appointment of an arbitrator there must have been a dispute between Servite and Multi which, or part of which, was substantially the same as that between Multi and Stent. Other esoteric legal arguments were advanced.

His Honour Judge James Fox-Andrews QC, official referee, held that the arbitrator was validly appointed under the proviso and had the requisite jurisdiction. In his judgment he made several important general points:

- 'It was an implied term of the sub-contract that the consent of the employer would be given to the exercise of the proviso within a reasonable time.'

On the facts, the necessary consent had been given and within a reasonable time.

- The sub-contract contained an implied term that the arbitrator appointed under the main contract would give his consent to acting also as arbitrator under [the sub-contract] within a reasonable time.'

That term was complied with.

- 'Where an appointment of an arbitrator is subject to any qualifications it is always desirable that the precise basis of the jurisdiction of the arbitrator should be recorded in writing and signed by the parties.'

The learned judge referred to an earlier case on the blue form arbitration clause before it was revised (*Higgis & Hill Building Ltd* v. *Campbell Denis Ltd* (1985)) which, emphasised that invoking the proviso is a matter for the judgment of the contractor, who as a party to both the main and sub-contract is best placed to know whether there are any matters in dispute between himself and the employer which are substantially the same as the dispute with the sub-contractor. He has to judge between the relative suitability of the courses open to him when an arbitrator is appointed under the sub-contract.

It is best if the main contractor's letter does spell out the precise respects in which the dispute or difference is substantially the same, but it is clear that a general assertion will do. One of the main advantages of arbitration is lost if jurisdictional and procedural points have to go before the courts.

The 1980 revisions of JCT and the related sub-contracts do away with some of the earlier obscurities. The current provisions suppose that once a dispute has been referred to an arbitrator under either the main or sub-contract, the arbitrator will have jurisdiction to deal with any related dispute, regardless of the wishes of the parties: see, for example, JCT 80, clause 41.2.1. This is subject to the proviso that either party may require the dispute or difference under the contract to be referred to a different arbitrator:

'if either of them reasonably considers that the arbitrator appointed to determine the related dispute is not appropriately qualified to determine the [present] dispute or difference ...'

Even so, the wording is still open to interpretation.

Sacking the arbitrator

Applications to revoke the authority of an arbitrator are very rare, although section 1 of the Arbitration Act 1950 is wide in its terms. It says:

'The authority of an arbitrator ... appointed by ... an arbitration agreement shall, unless the contrary intention is expressed in the agreement, be irrevocable *except by leave of the High Court ...*'

The difficulties thrown up by the Court of Appeal's decision in *Northern Regional Health Authority* v. *Derek Crouch Construction Co. Ltd* (1984), and the cases following it, have led to the suggestion that section 1 might not be a way out of the problem. In other words, if the parties are stuck with arbitration – and possibly with an appointed arbitrator not of their own choosing – could they now apply to the High Court for his authority to be revoked?

In a word, the answer is 'no', and the decided cases establish that granting leave under section 1 is an extreme remedy which is used only in the most unusual cases.

Mustill and Boyd's *Commercial Arbitration*, p. 471, suggests only four cases where the remedy is appropriate:

- serious and irreparable misconduct on the arbitrator's part;
- bias by the arbitrator;
- deficiencies in the arbitrator's capability or performance where there is no other remedy;
- situations in which justice demands that the arbitration proceedings should be temporarily halted or permanently brought to an end, and no other method of doing so is available to the court.

A recent case on section 1 is *Property Investments (Development) Ltd* v. *Byfield Building Services Ltd* (1985) and it aptly illustrates the limitations. It is a decision of Mr Justice Steyn who has wide experience of commercial arbitration, and emphasies that the parties cannot use a section 1 application merely on the grounds of convenience.

The plaintiff was extending its hotel in Daventry and employed the defendant as contractor. The work was to be done in phases and there were three separate contracts. The first two were in JCT 63 Form, with the arbitration clause, but the third contract was made by correspondence and did not incorporate JCT 63 terms. Disputes arose under all the contracts, and the defendant got an arbitrator appointed under the first two contracts, despite the plaintiff objecting to both the application and the appointment. In the case of the third contract, the defendant issued a writ claiming moneys allegedly due, and the employer put in a defence and counterclaim.

The appointed arbitrator held a preliminary meeting when the plaintiff challenged the validity of his appointment. The challenge was unsuccessful and the arbitrator decided that 'the matters in dispute are within the terms of' clause 35, but leaving the parties to have recourse to the court. Before Mr Justice Steyn it was conceded that the arbitrator was properly appointed and that the disputes under the first two contracts were within clause 35.

In support of their application to revoke the arbitrator's authority, the plaintiffs contended:

- if the arbitration were to continue there would be two separate actions in existence relating to the same building, the same workforce, the same parties, and the same witnesses:
- the disputes would be determined more quickly and cheaply if they were consolidated, which was not possible because of the arbitration proceedings;
- there was a related dispute about a supplier who was not a party to the arbitration agreement, which might well mean another set of proceedings.

Despite these seemingly compelling reasons and the wide language of section 1, Mr Justice Steyn refused to revoke the arbitrator's authority. The provision had to be considered in context and also in light of the wide range of remedies available to the court in connection with arbitration. A restrictive interpretation was to be adopted. 'The court ought only to revoke an arbitrator's authority in wholly exceptional circumstances,' he said, because the effect of an order under section 1 is to deprive a party of his contractual rights: *City Centre Properties Ltd* v. *Matthew Hall & Co. Ltd* (1969).

In his view, the court ought to revoke an arbitrator's authority only in *wholly exceptional circumstances* which fundamentally imperil the fair and proper functioning of the arbitration process.

Mr Justice Steyn was, indeed, of the view that in modern times the scope of section 1 should be restricted rather than widened, and of the four cases set out by Mustill and Boyd, felt that only the first two excited no controversy. He also summarised what he thought is the true commercial situation.

'The application is squarely based on grounds of convenience only and this will never warrant an order under section 1. Justice is an elusive concept in commercial relations but it certainly does not require that the court should deprive a party of his contractual right in relation to an agreed method of dispute resolution on grounds of convenience alone ... Plainly when two parties enter into a commercial contract containing an arbitration clause they know, or ought to know, that in multi-party disputes they will be unable to join other parties.'

These are robust words indeed, and will no doubt cause employers and contractors alike to consider whether or not they wish to have arbitration as the method of disputes settlement. It is the prescribed method in most standard form construction contracts, though I doubt whether the existence of the arbitration clause is actively in the parties' minds when they make the contract.

In the industry, of course, there is a major benefit in arbitration in that the arbitrator can review and revise the architect's decisions in a way that the court cannot do. This follows from the *Crouch* case, and must be taken to be the law unless and until we get a direct ruling from the Court of Appeal on the point. The difficulties cannot, however, be overcome by applying for the arbitrator's authority to be revoked.

Arbitration's Achilles' heel

The construction industry's chosen method of disputes settlement is arbitration, which has a number of advantages over litigation, one of them being that the hearing can be at the convenience of the parties and their witnesses. The growing popularity of arbitration in the industry is due in large measure to the Court of Appeal's holding in *Northern Regional Health Authority* v. *Derek Crouch Construction Ltd* (1984), holding that the courts do not possess the power to 'open up review and revise' decisions and certificates of the engineer or architect.

This self-denying ordinance does not mean that the courts cannot intervene in arbitration proceedings. Sections 22 and 23 of the Arbitration Act 1950 give the court broad powers to remit an award for reconsideration by the arbitrator, or to set the award aside or to remove the arbitrator from office where he 'has misconducted himself or the proceedings'.

'Misconduct' is an inappropriate term to describe what is in most cases a procedural lapse, as is well illustrated by the case of *Zermalt Holdings SA* v. *NuLife Upholstery Repairs Ltd* (1985), which was an application by a landlord to aside an award and to remove the arbitrator. The landlord argued that the reasons given in the award allegedly showed that the arbitrator had been influenced in his decision by matters which had not been raised by the parties and had not been put to the landlord for its comment. There was no criticism of the professional integrity, impartiality or competence of the arbitrator.

Mr Justice Bingham held that the landlord's case was made out. All matters likely to form the subject matter of the decision must be exposed for the comments and submissions of the parties. The arbitrator cannot in effect give evidence to himself. In his judgment, the judge made a number of important points. First, he emphasised as a matter of general approach, that the courts strive to uphold arbitrator's awards. They do not approach them with a meticulous legal eye and endeavour to pick holes. They strive to uphold awards and not strike them down.

He went on to describe the term 'misconduct' as an unfortunate one, since it 'gives the impression that some impropriety or breach of professional conduct or lack of integrity or incompetence is involved.' In 99 cases out of 100 an application to remove an arbitrator for misconduct involves nothing of the kind. 'It involves, usually, merely a procedural error of a kind that any arbitrator or magistrate or judge may be guilty of.'

Mr Justice Bingham went on to make an important statement about the distinction between the use of arbitrator's expertise in respect of general matters and the specific matters which must be put to the parties. It is of such importance that it deserves quotation at length.

'There is an unavoidable inclination to rely on one's own expertise and in respect of general matters that is not only not objectionable but is desirable and a very large part of the reason why an arbitrator with expert qualifications is chosen. Nevertheless, the rules of natural justice do require, even in an arbitration conducted by an expert, that matters which are likely to form the subject of the decision, in so far as they are specific matters, should be exposed for the comments and submissions of the parties.

If an arbitrator is impressed by a point that has never been raised by either side, then it is his duty to put it to them so that they have an opportunity to comment. If he feels that the proper approach is one that has not been explored ... then again it is his duty to give the parties a chance to comment. If he is to any extent relying on his own personal experience in a specific way then that is again something that he should mention so that it can be explored ...'

Construction industry arbitrators have taken note of this ruling and remember the case of *Fox* v. *P. G. Wellfair Ltd* (1981), where an award was set aside and the arbitrator was removed where the respondent did not appear and the arbitrator relied on his own technical knowledge to reject part of the claim without disclosing his views to the claimant during the proceedings.

Under the Arbitration Act 1979 an aggrieved party has a very limited right of appeal to the High Court on questions of law arising out of the award. Unless the parties consent, appeal will only lie with the consent of the court and leave to appeal will not be granted unless the court considers that the determination of the question of law could substantially affect the rights of one of the parties. Under JCT 80, clause 41.6, the parties agree at the time of entering into the contract that questions of law which arise in the course of the arbitration or out of the award shall be determined by the High Court under the provisions of the 1979 Act, but that 'agreement' is not necessarily effective under the Act.

Leave to appeal is given sparingly and the House of Lords in *The Nema* (1981) laid down strict guidelines, the effect of which is that the point of law must be of the widest possible application.

This strict approach carried a stage further by the Court of Appeal in *National Westminster Bank plc* v. *Arthur Young McClelland Moores & Co.* (1985). Section 1(7) of the 1979 Act says that no appeal lies to the Court of Appeal from a decision of the High Court on an appeal from an award unless the High Court or the Appeal Court gives leave and 'it is certified by the High Court that the question of law ... is one of general public importance ...'

In *Arthur Young* the High Court judge refused leave to appeal and also refused to certify that an important point of law was involved. The Court of Appeal ruled that the certificate can only be granted by the High Court, the Court of Appeal cannot certify. The whole point of the provision is to ensure finality and if the High Court says that there is no major legal point involved and refuses its certificate, there the matter ends.

The trend of all the recent cases is very clear. The parties must accept the decision of the arbitrator, whether it be right or wrong, unless the point of law involved is of general public importance.

And the emphasis is on *public* importance. There are many matters of importance to the construction industry and on which a ruling is desirable, but they are not matters of public importance. Unless the arbitrator makes a procedural lapse, effectively one is stuck with him and his decision.

Wise ways win costs awards

The cost of litigation or arbitration is often a significant factor in deciding whether to press ahead with a claim. Equally, the costs involved often exercise the minds of both claimants and respondents in arbitration proceedings in the context of 'offers to settle'.

The general situation about the award of costs in arbitration is quite clear. The arbitrator's powers to award costs and determine their amount are found in section 18 of the Arbitration Act 1950 which says in essence that they are in the arbitrator's discretion. But it is a discretion which must be exercised 'judicially' and in accordance with settled principles.

The starting point is the general rule that 'costs follow the event,' i.e. that in the ordinary way the successful party receives his costs. Not, I hasten to add, *all* his costs, because the usual basis is to award costs on what is called the 'standard basis', which means in practice that about two-thirds of the total bill is recoverable. This is described in many earlier judgments as the 'party and party' basis.

Complications can and do arise. A claim may succeed only in part, for example, or there may be a successful counterclaim. In that case, a possible order is that 'the claimant shall have the costs of the claim and the respondent shall have the costs of the counterclaim,' i.e. the arbitrator makes an apportionment. He may also do this in other cases, but there must be strong reasons for departing from the geneal rule.

A common case is where the claimant recovers less than was offered by the respondent in settlement, whether by way of an open or a sealed offer. In such a case, the arbitrator would award the respondent the costs incurred after the date of the offer. For a discussion of sealed offers, see John Parris, *Arbitration*, p. 148 *et seq.*

Mustill & Boyd, *Commercial Arbitration* (1982) p. 348, lists matters which have been held to justify a departure from the general rule. They include:

- gross exaggeration of the claim;
- unreasonable or obstructive conduct during the course of the

hearing which has lengthened the proceedings or increased the other party's costs;

- failure by the successful party on issues on which a large amount of time was spent;
- extravagance in the conduct of the hearing, e.g., employing an excess number of expert witnesses.

There are two common situations where guidance is often sought. The first is a dispute which involves, say, 15 separate heads of claim. The arbitrator finds in favour of the claimant under five heads, and on the remaining heads refuses to award any sum in respect of those heads, i.e., finds against the claimant. Is the arbitrator likely to apportion costs in such a case?

The answer to this question is that the arbitrator would probably apportion costs to the extent that they were necessarily incurred and no award should be made in respect of costs spent on pursuing obviously invalid claims.

Apportionment supposes, of course, that time *was* unnecessarily spent on the unsuccessful ten heads. If it were not then, in my view, there would be no reason to depart from the general rule.

'In the absence of special circumstances a successful litigant should receive his costs, and it is necessary to show some grounds of exercising the discretion of refusing an order which would give them to him, and the discretion must be exercised judicially': *Lewis* v. *Haverfordwest RDC* (1953).

A second situation is where there is a claim under a single head, and the arbitrator awards one-fifth of the sum claimed. In that case, unless there was a gross exaggeration of the claim (as in *Perry* v. *Stopher* (1959), where £54 was claimed and only £11 awarded!) there would be no reason to depart from the general rule. It is not possible to suggest a percentage of award to claim where the arbitrator would or should start to think about apportionment. As Mr Justice McNair observed in *Demolition & Construction Co. Ltd* v. *Kent River Board* (1963), there is no principle of law which compels the arbitrator to reflect the measure of success which one party or the other has achieved, in his award of costs, and it would be dangerous to suggest or think otherwise.

One method of dealing with this sort of problem is the device of the 'sealed offer' which is common in arbitration. But there are substantial objections to the practice which are usefully summarised in the Commercial Court Committee Report on Arbitration 1978, paragraphs 62–65. The better practice is to make an 'open offer' on terms that its existence and contents must not be disclosed to the arbitrator until he has reached his decision on liability when it will be drawn to his attention.

It should offer to pay costs up to the time of its receipt and state whether or not it includes interest. The letter should add that it is intended to have the effect of a payment into court under the Rules of the Supreme Court. Then, at the end of the hearing before the arbitrator, he should be asked to make an interim award on liability and amount and, without the offer being disclosed, he should be asked to defer consideration of costs until he has made his interim award. This procedure is commonly used and is suggested in Keating's *Building Contracts*, 4th . edition, pp. 274–275, although the practice was deprecated by the Commercial Court Committee!

In such a case, it is possible to apply the principle laid down in *Tramountana Armadora SA* v. *Atlantic Shipping Co. SA* (1978):

> 'If the claimant in the end has achieved no more than he would have achieved by accepting the offer, the continuance of the arbitration after that date has been a waste of time and money. *Prima facie*, the claimant should recover his costs up to the date of the offer and should be ordered to pay the respondent's costs after that date.'

The whole question of costs in arbitration causes much debate. The important point is that the arbitrator cannot act in an arbitrary way. The small size of the claim, for example, is not a ground on which a successful party may be deprived of his costs, and the ground rules are fairly clear.

Table of Cases

Index

statute-barred, 26, 36
sub-contractor's entitlement to,
    132–4
successful, 137–9
clauses
    alternative and conflicting, 19
    bankruptcy, 167
    direct payment, 127–9
    exclusion, 47, 48
    exemption, 45, 49
    extension of time, 100, 104
    fluctuation, 21, 74
    forfeiture, 43
    indemnity, 122
    interpretation, 30
    money claims, 139
    possession, 35
    price adjustment, 72
    retention of title, 166
    variation, 12
climatic conditions, 57
    exceptional, 102
    see also weather
common law, contractor's duty at,
    41
completion
    bonus for early, 106
    certificate of, 63
    date of, 67, 68
contra proferentem rule, 31, 105
contract (*see also* forms)
    acceptance of, 11
    additions to, 9
    anticipated, 5
    breach of, 49, 73, 161
        anticipatory, 163
    counter-offer, 12, 15
    documents, 30
    estimates as binding, 3
    exclusions and modifications, 39
    express, 32
    fixed price, 21, 35
    fluctuation clause, 21
    formation of, 13
    implied terms, 31, 39, 65
    interpretation of, 20, 29, 30

liability arising from tort or, 116
    modernisation, 34
    need for new, 11
    negotiation, for, 16
    pricing errors, 6
    reasonable implication of, 13
    reasonableness, 45, 46
    rectification of, 6
    refurbishment, 104
    repudiation of, 44, 60
    retrospective, 16
    retrospective breach, 114
    seal, under, 26, 36
    standard form, 19, 31, 35
    termination of, 60
    terms of, 11, 31, 39, 65
    time limits, 72
    to make a, 17
    'unfair terms', 45
    variation of, 61
    void, 23
contractors
    architect's duty of care towards,
        52–6
    chain of responsibility, 132
    completion programme, 67
    dual obligations of, 70
    due diligence and expedition,
        exercise of, 72
    duty of, to warn owner of
        defects, 59
    employment, determination of,
        169
    estimates by, 3
    reasonable remuneration of, 79
    recovery by, for damage, 38
    remedy of, against architect, 61
costs
    apportionment of, 191
    award of, 190
    offer of, 191–2
counter-offer, 12, 15

damage
    recovery by contractor for, 38
    special, 138